A computer program for
statistical power analysis
and confidence intervals

POWER AND
PRECISION

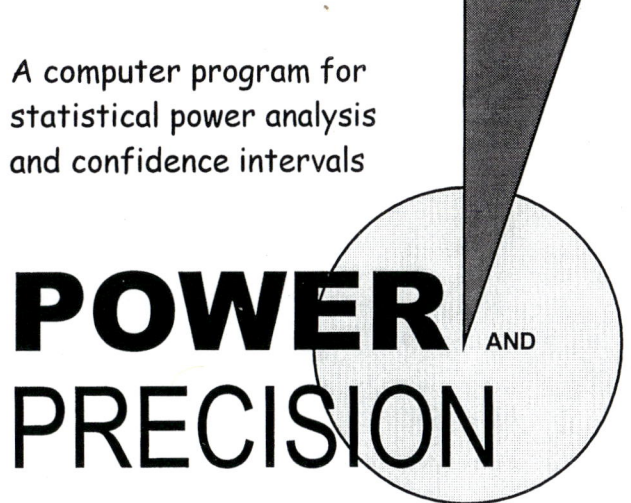

Michael Borenstein
Hannah Rothstein
Jacob Cohen

LAWRENCE ERLBAUM ASSOCIATES, PUBLISHERS
Mahwah, New Jersey

Distributed by Lawrence Erlbaum Associates, Inc.
All rights reserved. No part of this book may be reproduced
in any form, by photostat, microform, retrieval system, or any
other means, without the prior written permission of
the publisher.

Lawrence Erlbaum Associates, Inc., Publishers
10 Industrial Avenue
Mahwah, New Jersey 07430

ISBN 1-56321-198-X

Books published by Lawrence Erlbaum Associates are printed
on acid-free paper, and their bindings are chosen
for strength and durability.

Printed in the United States of America

10 9 8 7 6 5 4 3 2 1

For more information about Biostat® software products, please visit our WWW site at http://www.PowerAndPrecision.com or contact:

Biostat.
1421 Hudson Road
Teaneck, NJ 07666
Tel: (201) 692-8155, Fax (201) 692-9526
E-Mail Michael_Borenstein@PowerAndPrecision.com

Biostat is a registered trademark and the other product names are the trademarks of Biostat for its proprietary computer software. No material describing such software may be produced or distributed without the written permission of the owners of the trademark and license rights in the software and the copyrights in the published materials.

The SOFTWARE and documentation are provided with RESTRICTED RIGHTS. Use, duplication, or disclosure by the Government is subject to restrictions as set forth in subdivision (c)(1)(ii) of The Rights in Technical Data and Computer Software clause at 52.227-7013. Contractor/manufacturer is Biostat, 1421 Hudson Road, Teaneck, NJ 07666.

General notice: Other product names mentioned herein are used for identification purposes only and may be trademarks of their respective companies.

Windows 95 is a registered trademark of Microsoft Corporation.

Power and Precision™
Copyright © 1997 by Biostat
All rights reserved.
Printed in the United States of America.

No part of this publication may be reproduced, stored in a retrieval system, or transmitted, in any form or by any means, electronic, mechanical, photocopying, recording, or otherwise, without the prior written permission of the publisher

Preface

Power and Precision™ is a stand-alone software program that can be used by itself or as a tool to enhance any other statistical package. *Power and Precision* helps find an appropriate balance among effect size, sample size, the criterion required for significance (alpha), and power. Typically, when a study is being planned, either the effect size is known from previous research or an effect size of practical significance is specified. In addition, the user enters the desired alpha and power. The analysis then indicates the number of cases needed to attain the desired statistical power.

This program can also be used to obtain the number of cases needed for desired precision. In general, precision is a function of the confidence level required, the sample size, and the variance of the effect size.

In studies of either power or precision, the program produces graphs and tables, as well as written reports that are compatible with many word processing programs.

Compatibility

Power and Precision is designed to operate on computer systems running Windows 95, Windows NT 3.51, or Windows NT 4.0.

Customer Service

If you have any questions concerning your shipment or account, contact Biostat (see p. i.)

Technical Support

The services of Biostat Technical Support are available to registered customers of Biostat. Customers may contact Technical Support for assistance in using Biostat products or for installation help for one of the supported hardware environments. To reach Technical Support, see web site, e-mail, Fax and phone lines listed on p. i. Be prepared to identify yourself, your organization, and the serial number of your system.

Tell Us Your Thoughts

Your comments are important. Please send us a letter and let us know about your experiences using our products. Write to Biostat, 1421 Hudson Road, Teaneck, NJ 07666.

Acknowledgments

We gratefully acknowledge the support of the National Institute of Mental Health (NIMH), which made the development of this software possible through the Small Business Innovation Research Program. In particular, we thank the following people at NIMH for their enthusiasm, help, and encouragement: Gloria Levin, scientific review administrator; Michael Huerta, associate director, Division of Neuroscience and Behavioral Research; Mary Curvey, grants program specialist, Division of Neuroscience and Behavioral Research; and James Moynihan, former associate director (now retired).

Don Rubin, Larry Hedges, John Barnard, and Kim Meier helped in the development of specific algorithms.

The algorithm for non-central F was developed by Barry W. Brown, James Lovato, and Kathy Russell of the Department of Biomathematics, University of Texas, M.D. Anderson Cancer Center, Houston, Texas (through a grant from the National Cancer Institute) and is used with their kind permission. Some of the program's procedures for exact tests of proportions were adapted from code originally written by Gerard E. Dallal and used with his kind permission. The program also employs algorithms under license from IMSL, Inc.

John Kane generously allowed the senior author to take a partial sabbatical from his work at Hillside Hospital to work on this program.

Michael Borenstein
Hillside Hospital,
Albert Einstein College of Medicine,
and Biostatistical Programming Associates

Hannah Rothstein
Baruch College, City University of New York

Jacob Cohen
New York University

Contents

1 The Sixty-Second Tour   1
Selecting a Procedure   1
Navigating within the Program   2

2 Overview of Power and Precision   5
Power Analysis   5
  Effect Size   5
  Alpha   7
  Tails   9
  Sample Size   9
  Power   10
  Ethical Issues   10
  Additional Reading   11
Precision   11
  Sample Size   12
  Confidence Level   12
  Tails   13
  Variance of the Effect Size   14
  Effect Size   14
  Planning for Precision   15
  Tolerance Intervals   16
  Additional Reading   17
Significance Testing versus Effect Size Estimation   17
  Additional Reading   18

3 The Main Screen   19
Computing Power Interactively   19
Toolbar   20
Interactive Guide   21

Summary Panel 22
    Effect Size Conventions 22
    Sensitivity Analysis 23
    Alpha, Confidence Level, and Tails 26
    Finding the Number of Cases Automatically 28
    Modifying the Default Value of Power 29
    Printing the Main Screen 29
    Copying to the Clipboard 30
    Saving and Restoring Files 30

4 Tables 31

Table Style 32
    Power and Precision 32
    Power Only 32

Effect Size 35

Alpha and/or Tails 36
    Varying Effect Size and Alpha 37

Sample Size 38

Computational Options 39

Modifying Titles 39

Printing 39

Copying to the Clipboard 39

Saving to a File 40

Generating a Graph 40

5 Graphs 41

Customizing Graphs 45
    Modifying Content 45
    Formatting Titles, Legends, and Colors 46

Printing Graphs 47

Copying Graphs to the Clipboard 47

Saving Graphs to Disk 48

6 Reports 49

Printing Reports 51

Copying Reports to the Clipboard 52

Saving Reports 52

7 T-Test for One Group 53

Selecting the Procedure 53

Application 54
- Effect Size 54
- Alpha, Confidence Level, and Tails 55
- Sample Size 55
- Tolerance Interval 55
- Computational Options for Power 56
- Computational Options for Precision 56
- Options for Screen Display 56

Example 57

8 Paired T-Test 59

Selecting the Procedure 59

Application 60
- Effect Size 61
- Alpha, Confidence Level, and Tails 62
- Sample Size 63
- Tolerance Interval 63
- Computational Options for Power 64
- Computational Options for Precision 64
- Options for Screen Display 64

Example 64

9 T-Test for Independent Groups 67

Selecting the Procedure 67

Application 68
- Effect Size 68
- Alpha, Confidence Level, and Tails 69
- Sample Size 70

Tolerance Interval   70
Computational Options for Power   71
Computational Options for Precision   71
Options for Screen Display   71

Example   72

10 Proportions in One Sample   75

Selecting the Procedure   75

Application   76
Effect Size   76
Alpha, Confidence Level, and Tails   77
Sample Size   77
Computational Options for Power   77
Computational Options for Precision   78

Example   78

11 Proportions in Two Independent Groups   81

Selecting the Procedure   81

Application   82
Effect Size   82
Alpha, Confidence Level, and Tails   83
Sample Size   83
Computational Options for Power   83
Computational Options for Precision   84
Options for Screen Display   84

Example   85

12 Paired Proportions   87

Selecting the Procedure   87

Application   88
Effect Size   88
Alpha and Tails   89
Sample Size   89
Computational Options for Power   89

Example   90

13 Sign Test 93
Selecting the Procedure 93

Application 94
   Effect Size 94
   Alpha and Tails 94
   Sample Size 95
   Computational Options for Power 95

Example 95

14 K x C Crosstabulation 97
Selecting the Procedure 97

Application 98
   Effect Size 98
   Alpha 99
   Sample Size 99
   Computational Options for Power 100

Example 100

15 Correlation—One Group 103
Selecting the Procedure 103

Application 104
   Effect Size 104

Alpha, Confidence Level, and Tails 104
   Sample Size 105
   Computational Options for Power 105

Example 1 105

Example 2 107

16 Correlation—Two groups 109
Selecting the Procedure 109

Application 110
   Effect Size 110

Alpha, Confidence Level, and Tails   111
   Sample Size   111
   Computational Options for Power   111

Example   111

17 Analysis of Variance/Covariance (Oneway)   113

Selecting the Procedure   113

Application   113
   Effect Size   114
   Entering the Effect Size (f) for Oneway ANOVA   115
   Correspondence between the Four Approaches   120
   Effect Size Updated Automatically   121
   Alpha   121
   Sample Size   121

Example 1   122

Oneway Analysis of Covariance   123

Example 2   124

18 Analysis of Variance/Covariance (Factorial)   127

Selecting the Procedure   127

Application   127
   Effect Size   128
   Entering the Effect Size (f) for Factorial ANOVA   129
   Correspondence between the Four Approaches   134
   Effect Size Updated Automatically   135
   Alpha   135
   Sample Size   135

Example 1   136

Factorial Analysis of Covariance   138

Example 2   139
   Generate Table   141
   Generate Graph   142

19 Multiple Regression 143

Selecting the Procedure 143

Application 144
- The Designated Set 145
- Effect Size 146
- Alpha and Tails 146
- Sample Size 146
- Computational Options for Power 146
- Options for Study Design 148

Example 1 149
- One Set of Variables 149

Example 2 151
- Set of Covariates Followed by Set of Interest 151

Example 3 153
- Two Sets of Variables and Their Interaction 153

Example 4 155
- Polynomial Regression 155

Example 5 157
- One Set of Covariates, Two Sets of Variables, and Interactions 157

20 General Case 163

General Case (Non-Central T) 164

General Case (Non-Central F) 167

General Case (Non-Central Chi-Square) 170

Printing the General Case Panel 172

Copying Data to the Clipboard 172

Appendix A
Installation 173

Requirements 173

Installing 173

Uninstalling 173

Troubleshooting 173

Note 174

Appendix B
Troubleshooting 175

Appendix C
Computational Algorithms for Power 177

Computation of Power 177

T-Test (One-Sample and Paired) with Estimated Variance 177

T-Test (Two-Sample) with Estimated Variance 178

Z-Test (One-Sample and Paired) with Known Variance 179

Z-Test (Two-Sample) with Known Variance 180

Single Proportion versus a Constant 182

Two Independent Proportions 183
  Two Proportions: Arcsin Method 183
  Two Proportions: Normal Approximations 184
  Two Proportions—Chi-Square Test 185
  Two Proportions: Fisher's Exact Test 185
  One-Tailed versus Two-Tailed Tests 186

McNemar Test of Paired Proportions 186

Sign Test 187

K x C Crosstabulation 187
  Computing Effect Size 187

Correlations—One-Sample 188
  Pearson Correlation—One-Sample versus Zero 188
  One-Sample versus Constant Other than Zero 188

Correlations—Two-Sample   189
Analysis of Variance   190
   Computing the Effect Size (f)   190
Analysis of Covariance   192
Multiple Regression   192
   Definition of Terms   192
   Model 1 versus Model 2 Error   193
   Computation of Power   193

Appendix D
Computational Algorithms for Precision   195

T-Test for Means (One-Sample) with Estimated Variance   195
T-Test for Means (Two-Sample) with Estimated Variance   196
Z-Test for Means (One-Sample)   197
Z-Test for Means (Two-Sample)   198
One Proportion—Normal Approximation   199
One Proportion—Exact (Binomial) Formula   200
Two Independent Proportions—Normal Approximations   201
   Rate Difference   201
   Odds Ratio   202
   Relative Risk   202
Two Independent Proportions—Cornfield Method   202
Correlations (One-Sample)   203

Glossary   205

Bibliography   209

Index   215

1 The Sixty-Second Tour

Selecting a Procedure

To display the available procedures, select *New analysis* from the File menu.

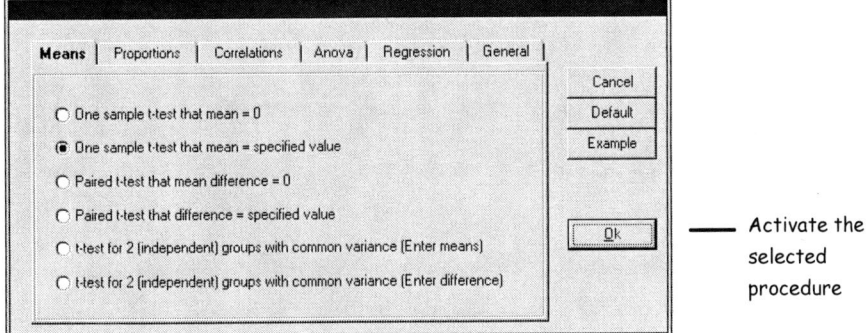

Activate the selected procedure

Procedures are grouped into the following categories:
- Means (t-tests and z-tests for one group and for two groups)
- Proportions (tests of single proportions, of two proportions for independent or matched groups, and for a K × C crosstabulation
- Correlations (one- and two-group correlations)
- ANOVA (analysis of variance and covariance—oneway or factorial)
- Multiple regression (for any number of sets, for increment or cumulative R^2)
- General case (allows user to specify the non-centrality parameter directly)

To see an example, click *Example*. To activate a procedure, click *OK*.

Navigating within the Program

▶ Enter data on main panel.

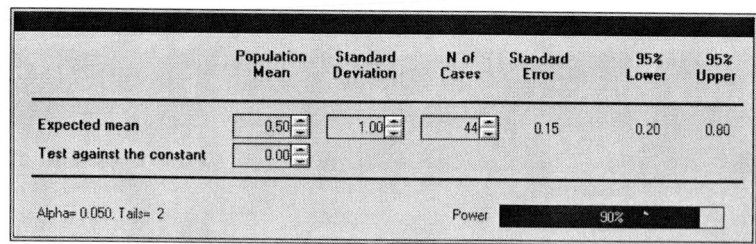

Enter basic data for effect size, alpha, confidence level, and sample size. The program will display power and precision and will update them continuously as study parameters are set and modified.

▶ Click the *Report* icon to generate a report incorporating the data from the main screen..

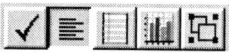

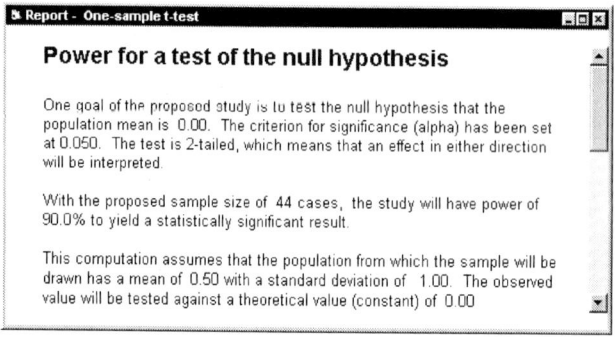

▶ Click the *Tables* icon to generate a table showing how power or precision will vary as a function of sample size. The table is based on the data from the main screen..

		Raw Mean = 0.5			Std Mean = 0.5		
N [1]	Power	Standard Error	95% Lower	95% Upper	Standard Error	95% Lower	95% Upper
10	.293	.316	-.170	1.170	.316	-.170	1.170
12	.353	.289	-.102	1.102	.289	-.102	1.102
14	.410	.267	-.052	1.052	.267	-.052	1.052
16	.465	.250	-.012	1.012	.250	-.012	1.012
18	.516	.236	.020	.980	.236	.020	.980
20	.565	.224	.046	.954	.224	.046	.954
22	.609	.213	.069	.931	.213	.069	.931
24	.650	.204	.088	.912	.204	.088	.912
26	.688	.196	.105	.895	.196	.105	.895
28	.723	.189	.121	.879	.189	.121	.879
30	.754	.183	.134	.866	.183	.134	.866

Power and Precision as a Function of Sample Size
Mean in one sample versus a constant

▶ Click the *Graphs* icon to generate a graph incorporating the data from the main screen.

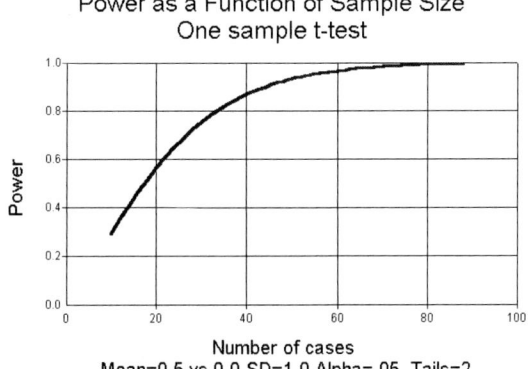

Power as a Function of Sample Size
One sample t-test
Number of cases
Mean=0.5 vs 0.0 SD=1.0 Alpha=.05 Tails=2

▶ The program includes various tools that assist the user to set an effect size or to find the sample size required for power. Use these tools to find an appropriate balance between alpha and beta and/or to ensure adequate precision.

2 Overview of Power and Precision

Power Analysis

Traditionally, data collected in a research study is submitted to a significance test to assess the viability of the null hypothesis. The p-value, provided by the significance test and used to reject the null hypothesis, is a function of three factors: size of the observed effect, sample size, and the criterion required for significance (alpha).

A power analysis, executed when the study is being planned, is used to anticipate the likelihood that the study will yield a significant effect and is based on the same factors as the significance test itself. Specifically, the larger the effect size used in the power analysis, the larger the sample size, and/or the larger (more liberal) the criterion required for significance (alpha), the higher the expectation that the study will yield a statistically significant effect.

These three factors, together with power, form a closed system—once any three are established, the fourth is completely determined. The goal of a power analysis is to find an appropriate balance among these factors by taking into account the substantive goals of the study, and the resources available to the researcher.

Effect Size

The term *effect size* refers to the magnitude of the effect under the alternate hypothesis. The nature of the effect size will vary from one statistical procedure to the next (it could be the difference in cure rates, or a standardized mean difference, or a correlation coefficient), but its function in power analysis is the same in all procedures.

The effect size should represent the smallest effect that would be of clinical or substantive significance, and for this reason, it will vary from one study to the next. In clinical trials, for example, the selection of an effect size might take into account the severity of the illness being treated (a treatment effect that reduces mortality by 1% might be clinically important, while a treatment effect that reduces transient asthma by 20% may be of little interest). It might take into account the existence of alternate treatments. (If alternate treatments exist, a new treatment would need to surpass these other treatments to be important.) It might also take into account the treatment's cost and side effects. (A treatment that carried these burdens would be adopted only if the treatment effect was very substantial.)

Power analysis gives power for a specific effect size. For example, the researcher might report that if the treatment increases the recovery rate by 15 percentage points, the study will have power of 80% to yield a significant effect. For the same sample size and alpha, if the treatment effect is less than 15 percentage points, then the power will be less than 80%. If the true effect size exceeds 15 percentage points, then power will exceed 80%.

While one might be tempted to set the "clinically significant effect" at a small value to ensure high power for even a small effect, this determination cannot be made in isolation. The selection of an effect size reflects the need for balance between the size of the effect that we can detect and resources available for the study.

Figure 2.1 Power as a function of effect size and N

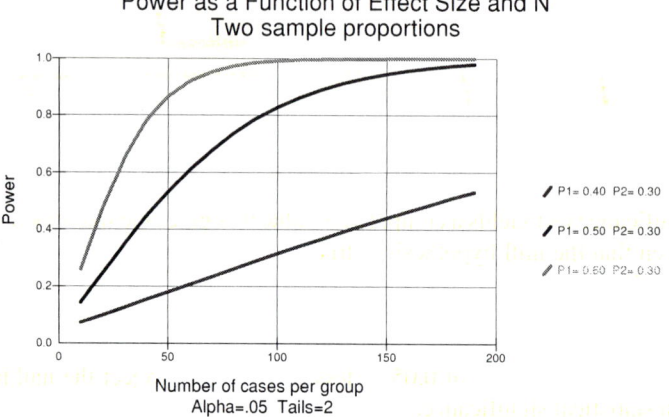

Small effects will require a larger investment of resources than large effects. Figure 2.1 shows power as a function of sample size for three levels of effect size (assuming that alpha, two-tailed, is set at 0.05). For the smallest effect (30% versus 40%), we would need a sample of 356 per group to yield power of 80% (not shown on the graph). For the intermediate effect (30% versus 50%), we would need a sample of 93 per group to yield this level of power. For the largest effect size (30% versus 60%), we would need a sample of 42 per group to yield power of 80%. We may decide that for our purposes, it would make sense to enroll 93 per group to detect the intermediate effect but inappropriate to enroll 356 patients per group to detect the smallest effect.

The *true* (population) effect size is not known. While the effect size used for the power analysis is assumed to reflect the population effect size, the power analysis is more appropriately expressed as, "*If* the true effect is this large, power would be ...," rather than, "The true effect is this large, and therefore power is ...."

This distinction is an important one. Researchers sometimes assume that a power analysis cannot be performed in the absence of pilot data. In fact, it is usually possible to perform a power analysis based entirely on a logical assessment of what constitutes a clinically (or theoretically) important effect. Indeed, while the effect observed in prior studies might help to provide an *estimate* of the true effect, it is not likely to be the *true* effect in the population—if we knew that the effect size in these studies was accurate, there would be no need to run the new study.

Since the effect size used in power analysis is not the *true* population value, the researcher may decide to present a range of power estimates. For example (assuming that N = 93 per group and alpha = 0,05, two-tailed), the researcher may state that the study will have power of 80% to detect a treatment effect of 20 points (30% versus 50%) and power of 99% to detect a treatment effect of 30 points (30% versus 60%).

Cohen has suggested conventional values for small, medium, and large effects in the social sciences. The researcher may want to use these values as a kind of reality check to ensure that the values that he or she has specified make sense relative to these anchors. The program also allows the user to work directly with one of the conventional values rather than specifying an effect size, but it is preferable to specify an effect based on the criteria outlined above, rather than relying on conventions.

Alpha

The significance test yields a computed p-value that gives the likelihood of the study effect, given that the null hypothesis is true. For example, a p-value of 0.02 means that, assuming that the treatment has a null effect, and given the sample size, an effect as large as the observed effect would be seen in only 2% of studies.

The p-value obtained in the study is evaluated against the criterion, alpha. If alpha is set at 0.05, then a p-value of 0.05 or less is required to reject the null hypothesis and establish statistical significance.

If a treatment really is effective and the study succeeds in rejecting the nil hypothesis, or if a treatment really has no effect and the study fails to reject the nil hypothesis, the study's result is correct. A type 1 error is said to occur if there is a nil effect but we mistakenly reject the null. A type 2 error is said to occur if the treatment is effective but we fail to reject the nil hypothesis.

Note: The *null hypothesis* is the hypothesis to be nullified. When the null hypothesis posits a nil effect (for example, a mean difference of 0), the term *nil hypothesis* is used.

Assuming that the null hypothesis is true and alpha is set at 0.05, we would expect a type I error to occur in 5% of all studies—the type I error rate is equal to alpha. Assuming that the null hypothesis is false (and the true effect is given by the effect size used in computing power), we would expect a type 2 error to occur in the proportion of studies denoted by one minus power, and this error rate is known as beta.

If our only concern in study design were to prevent a type 1 error, it would make sense to set alpha as conservatively as possible (for example, at 0.001). However, alpha does not operate in isolation. For a given effect size and sample size, as alpha is decreased, power is also decreased. By moving alpha from, say, 0.10 toward 0.01, we reduce the likelihood of a type 1 error but increase the likelihood of a type 2 error.

Figure 2.2   Power as a function of alpha and N

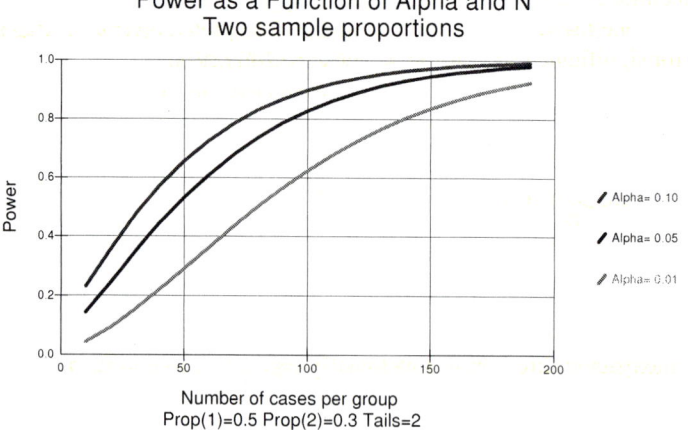

Figure 2.2 shows power as a function of sample size for three levels of alpha (assuming an effect size of 30% versus 50%, which is the intermediate effect size in the previous figure). For the most stringent alpha (0.01), an N of 139 per group is required for power of 0.80. For alpha of 0.05, an N of 93 per group is required. For alpha of 0.10, an N of 74 per group is required.

Traditionally, researchers in some fields have accepted the notion that alpha should be set at 0.05 and power at 80% (corresponding to a type 2 error rate and beta of 0.20). This notion implies that a type 1 error is four times as harmful as a type 2 error (the ratio of alpha to beta is 0.05 to 0.20), which provides a *general* standard in a specific application. However, the researcher must strike a balance between alpha and beta appropriate to the specific issues. For example, if the study will be used to screen a new drug for further testing, we might want to set alpha at 0.20 and power at 95% to ensure that a potentially useful drug is not overlooked. On the other hand, if we were working with a drug that carried the risk of side effects and the study goal was to obtain FDA approval for use, we might want to set alpha at 0.01, while keeping power at 95%.

Tails

The significance test is always defined as either one-tailed or two-tailed. A two-tailed test is a test that will be interpreted if the effect meets the criterion for significance and falls in either direction. A two-tailed test is appropriate for the vast majority of research studies. A one-tailed test is a test that will be interpreted only if the effect meets the criterion for significance and falls in the observed direction (that is, the treatment *improves* the cure rate) and is appropriate only for a specific type of research question.

Cohen gives the following example of a one-tailed test. An assembly line is currently using a particular process (A). We are planning to evaluate an alternate process (B), which would be expensive to implement but could yield substantial savings if it works as expected. The test has three possible outcomes: process A is better, there is no difference between the two, or process B is better. However, for our purposes, outcomes 1 and 2 are functionally equivalent, since either would lead us to maintain the status quo. In other words, we have no need to distinguish between outcomes 1 and 2.

A one-tailed test should be used only in a study in which, as in this example, an effect in the unexpected direction is functionally equivalent to no effect. It is *not* appropriate to use a one-tailed test simply because one is able to specify the expected direction of the effect prior to running the study. In medicine, for example, we typically expect that the new procedure will *improve* the cure rate, but a finding that it decreases the cure rate would still be important, since it would demonstrate a possible flaw in the underlying theory.

For a given effect size, sample size, and alpha, a one-tailed test is more powerful than a two-tailed test (a one-tailed test with alpha set at 0.05 has approximately the same power as a two-tailed test with alpha set at 0.10). However, the number of tails should be set based on the substantive issue of whether an effect in the reverse direction will be meaningful. In general, it would not be appropriate to run a test as one-tailed rather than two-tailed as a means of increasing power. (Power is higher for the one-tailed test only under the assumption that the observed effect falls in the expected direction. When the test is one-tailed, power for an effect in the reverse direction is nil).

Sample Size

For any given effect size and alpha, increasing the sample size will increase the power (ignoring for the moment the special case where power for a test of proportions is computed using exact methods). As is true of effect size and alpha, sample size cannot be viewed in isolation but rather as one element in a complex balancing act. In some studies, it might be important to detect even a small effect while maintaining high power. In such a case, it might be appropriate to enroll many thousands of patients (as was done in the physicians' study that found a relationship between aspirin use and cardiovascular events).

Typically, though, the number of available cases is limited. The researcher might need to find the largest N that can be enrolled and work backward from there to find an appropriate balance between alpha and beta. She may need to forgo the possibility of finding a small effect and acknowledge that power will be adequate for a large effect only.

Note. For studies that involve two groups, power is generally maximized when the total number of subjects is divided equally between two groups. When the number of cases in the two groups is not equal, the "effective N" for computing power falls closer to the smaller sample size than the larger one.

Power

Power is the fourth element in this closed system. Given an effect size, alpha, and sample size, power is determined. As a general standard, power should be set at 80%. However, for any given research, the appropriate level of power should be decided on a case-by-case basis, taking into account the potential harm of a type 1 error, the determination of a clinically important effect, and the potential sample size, as well as the importance of identifying an effect, should one exist.

Ethical Issues

Some studies involve putting patients at risk. At one extreme, the risk might be limited to loss of time spent completing a questionnaire. At the other extreme, the risk might involve the use of an ineffective treatment for a potentially fatal disease. These issues are clearly beyond the scope of this discussion, but one point should be made here.

Ethical issues play a role in power analysis. If a study to test a new drug will have adequate power with a sample of 100 patients, then it would be inappropriate to use a sample of 200 patients, since the second 100 are being put at risk unnecessarily. At the same time, if the study requires 200 patients in order to yield adequate power, it would be inappropriate to use only 100. These 100 patients may consent to take part in the study on the assumption that the study will yield useful results. If the study is underpowered, then the 100 patients have been put at risk for no reason.

Of course, the actual decision-making process is complex. One can argue about whether adequate power for the study is 80%, 90%, or 99%. One can argue about whether power should be set based on an improvement of 10 points, 20 points, or 30 points. One can argue about the appropriate balance between alpha and beta. In addition, the sample size should take account of precision as well as power (see "Precision," below). The point here is that these kinds of issues need to be addressed explicitly as part of the decision-making process.

The Null Hypothesis versus the Nil Hypothesis

Power analysis focuses on the study's potential for rejecting the null hypothesis. In most cases, the null hypothesis is the null hypothesis of no effect (also known as the nil hypothesis). For example, the researcher is testing a null hypothesis that the change in score from time 1 to time 2 is 0. In some studies, however, the researcher might attempt to disprove a null hypothesis other than the nil. For example, if the researcher claims that the intervention boosts the scores by 20 points or more, the impact of this claim is to change the effect size.

Additional Reading

The bibliography includes a number of references that offer a more comprehensive treatment of power. In particular, see Borenstein (1994B, 1997), Cohen (1965, 1988, 1992, 1994), Feinstein (1975), and Kraemer (1987).

Precision

The discussion to this point has focused on power analysis, which is an appropriate precursor to a test of significance. If the researcher is designing a study to test the null hypothesis, then the study design should ensure, to a high degree of certainty, that the study will be able to provide an adequate (that is, powerful) test of the null hypothesis.

The study may be designed with another goal as well. In addition to (or instead of) testing the null hypothesis, the researcher might use the study to estimate the magnitude of the effect—to report, for example, that the treatment increases the cure rate by 10 points, 20 points, or 30 points. In this case, study planning would focus not on the study's ability to reject the null hypothesis but rather on the precision with which it will allow us to estimate the magnitude of the effect.

Assume, for example, that we are planning to compare the response rates for treatments and anticipate that these rates will differ from each other by about 20 percentage points. We would like to be able to report the rate difference with a precision of plus or minus 10 percentage points.

The precision with which we will be able to report the rate difference is a function of the confidence level required, the sample size, and the variance of the outcome index. Except in the indirect manner discussed below, it is not affected by the effect size.

Sample Size

The confidence interval represents the precision with which we are able to report the effect size, and the larger the sample, the more precise the estimate. As a practical matter, sample size is the dominant factor in determining the precision.

Figure 2.3 Precision for a rate difference (95% confidence interval), effect size 20 points

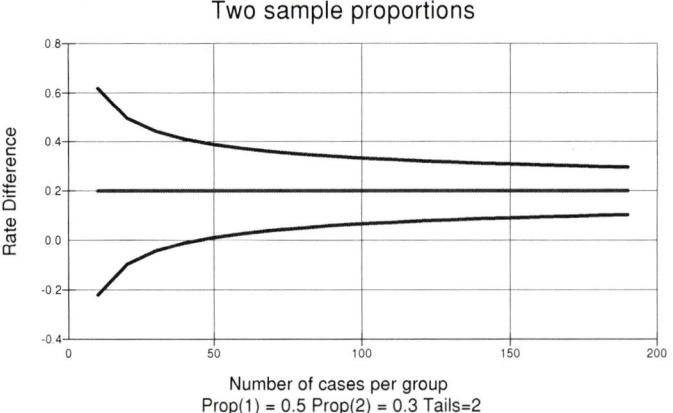

Figure 2.3 shows precision for a rate difference as a function of sample size. This figure is based on the same rates used in the power analysis (30% versus 50%). With N = 50 per group, the effect would be reported as 20 points, with a 95% confidence interval of plus or minus 19 points (01 to 39 points). With N= 100 per group, the effect would be reported as 20 points, with a 95% confidence interval of plus or minus 13 points (7 to 33). With N = 200 per group, the effect would be reported as 20 points, with a 95% confidence interval of plus or minus 9 points (11 to 29).

Note. For studies that involve two groups, precision is maximized when the subjects are divided equally between the two groups (this statement applies to the procedures included in this program). When the number of cases in the two groups is uneven, the "effective N" for computing precision falls closer to the smaller sample size than the larger one.

Confidence Level

The confidence level is an index of certainty. For example, with N= 93 per group, we might report that the treatment improves the response rate by 20 percentage points, with a 95% confidence interval of plus or minus 13 points (7 to 33). This means that in 95%

of all possible studies, the confidence interval computed in this manner will include the true effect. The confidence level is typically set in the range of 99% to 80%.

Figure 2.4 Precision for a rate difference (80% confidence interval)

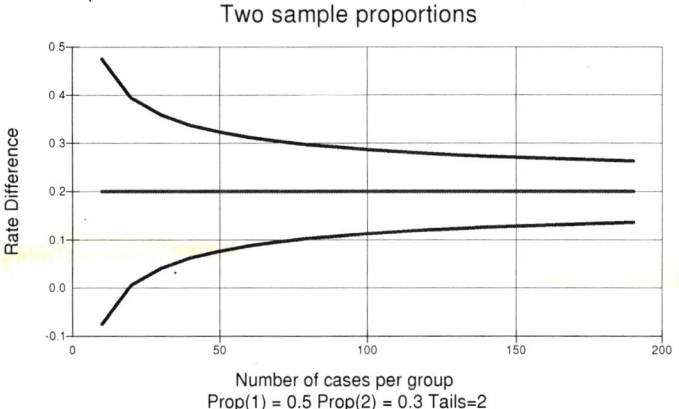

The 95% confidence interval will be wider than the 90% interval, which in turn will be wider than the 80% interval. For example, compare Figure 2.4, which shows the expected value of the 80% confidence interval, with Figure 2.3, which is based on the 95% confidence interval. With a sample of 100 cases per group, the 80% confidence interval is plus or minus 9 points (11 to 29), while the 95% confidence interval is plus or minus 13 points (7 to 34).

The researcher may decide to report the confidence interval for more than one level of confidence. For example, he may report that the treatment improves the cure rate by 10 points (80% confidence interval 11 to 29, and 95% confidence interval 7 to 34). It has also been suggested that the researcher use a graph to report the full continuum of confidence intervals as a function of confidence levels. (See Poole, 1987a,b,c; Walker, 1986a,b.)

Tails

The researcher may decide to compute two-tailed or one-tailed bounds for the confidence interval. A two-tailed confidence interval extends from some finite value below the observed effect to another finite value above the observed effect. A one-tailed confidence "interval" extends from minus infinity to some value above the observed effect, or from some value below the observed effect to plus infinity (the logic of the procedure may impose a limit other than infinity, such as 0 and 1, for proportions). A one-tailed confidence interval might be used if we were concerned with effects in only one direction. For example, we might report that a drug increases the remission rate by 20 points, with a 95% lower limit of 15 points (the upper limit is of no interest).

For any given sample size, dispersion, and confidence level, a one-tailed confidence interval is narrower than a two-tailed interval in the sense that the distance from the observed effect to the computed boundary is smaller for the one-tailed interval (the one-tailed case is not really an interval, since it has only one boundary). As was the case with power analysis, however, the decision to work with a one-tailed procedure rather than a two-tailed procedure should be made on substantive grounds rather than as a means for yielding a more precise estimate of the effect size.

Variance of the Effect Size

The third element determining precision is the dispersion of the effect size index. For t-tests, dispersion is indexed by the standard deviation of the group means. If we will be reporting precision using the metric of the original scores, then precision will vary as a function of the standard deviation. (If we will be reporting precision using a standard index, then the standard deviation is assumed to be 1.0, thus the standard deviation of the original metric is irrelevant.) For tests of proportions, the variance of the index is a function of the proportions. Variance is highest for proportions near 0.50 and lower for proportions near 0.0 or 1.0. As a practical matter, variance is fairly stable until proportions fall below 0.10 or above 0.90. For tests of correlations, the variance of the index is a function of the correlation. Variance is highest when the correlation is 0.

Effect Size

Effect size, which is a primary factor in computation of power, has little, if any, impact in determining precision. In the example, we would report a 20-point effect, with a 95% confidence interval of plus or minus 13 points. A 30-point effect would similarly be reported, with a 95% confidence interval of plus or minus 13 points.

Compare Figure 2.5, which is based on an effect size of 30 points, with Figure 2.3, which is based on an affect size of 20 points. The width of the interval is virtually identical in the two figures; in Figure 2.5, the interval is simply shifted north by 10 percentage points.

Figure 2.5 Precision for a rate difference (95% confidence interval), effect size 30 points

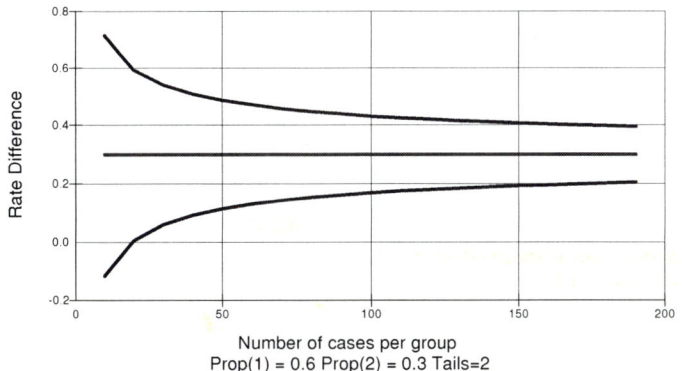

While effect size plays no direct role in precision, it may be related to precision indirectly. Specifically, for procedures that work with mean differences, the effect size is a function of the mean difference and also the standard deviation within groups. The former has no impact on precision; the latter affects both effect size and precision (a smaller standard deviation yields higher power and better precision in the raw metric). For procedures that work with proportions or correlations, the absolute value of the proportion or correlation affects the index's variance, which in turn may have an impact on precision.

Planning for Precision

The process of planning for precision has some obvious parallels to planning for power, but the two processes are not identical and, in most cases, will lead to very different estimates for sample size. The program displays the expected value of the precision for a given sample size and confidence level. In Figure 2.6, the user has entered data for effect size and found that a sample of 124 per group will yield power of 90% and precision (95% confidence interval) of plus or minus 12 points.

Figure 2.6 Proportions, 2 X 2 independent samples

	Response Rate	N Per Group	Standard Error	95% Lower	95% Upper
New treatment	0.50	124			
Standard treatment	0.30	124			
Rate Difference	0.20	248	0.06	0.08	0.32

Alpha= 0.05, Tails= 2   Power   90%

Typically, the user will enter data for effect size and sample size. The program immediately displays both power and precision for the given values. Changes to the effect size will affect power (and may have an incidental effect on precision). Changes to sample size will affect both power and precision. Changes to alpha will affect power, while changes to the confidence level will affect precision. Defining the test as one-tailed or two-tailed will affect both power and precision.

Tolerance Intervals

The confidence interval width displayed for t-tests is the median interval width. (Assuming the population standard deviation is correct, the confidence interval will be narrower than the displayed value in half of the samples and wider in half of the samples). The width displayed for single proportions using exact methods is the expected value (that is, the mean width expected over an infinite number of samples). For other procedures where the program displays a confidence interval, the width shown is an approximate value: It is the width that would be reported if the sample were to yield a statistic (i.e. a proportion or correlation) close to the value entered by the user for the effect size.

For many applications, especially when the sample size is large, these values will prove accurate enough for planning purposes. Note, however, that for any single study, the precision will vary somewhat from the displayed value. For t-tests, on the assumption that the population standard deviation is 10, the sample standard deviation will typically be smaller or greater than 10, yielding a narrower or wider confidence interval. Analogous issues exist for tests of proportions or correlations.

For t-tests, the researcher who requires more definitive information about the confidence interval may want to compute tolerance intervals—that is, the likelihood that the confidence interval will be no wider than some specific value. In this program, the 50% tolerance interval (corresponding to the median value) is displayed as a matter of course. The 80% (or other user-specified) tolerance interval is an option enabled from the View menu. For example, the researcher might report that in 50% of all studies, the mean would be reported with a 95% confidence interval no wider than 9 points, and in 80%

of all studies, the mean would be reported with a 95% confidence interval no wider than 10 points.

Note. The confidence interval displayed by the program is intended for anticipating the width of the confidence interval while planning a study, and not for computing the confidence interval after a study is completed. The computational algorithm used for t-tests includes an adjustment for the sampling distribution of the standard deviation that is appropriate for planning but not for analysis. The computational algorithms used for tests of proportions or a single correlation may be used for analysis as well.

Additional Reading

The bibliography includes a number of references that offer a more comprehensive treatment of precision. In particular, see Borenstein, 1994; Cohen, 1994; Bristol, 1989; Gardner and Altman, 1986; and Hahn and Meeker, 1991.

Significance Testing versus Effect Size Estimation

The two approaches outlined here—testing the null hypothesis of no effect and estimating the size of the effect—are closely connected. A study that yields a p-value of precisely 0.05 will yield a 95% confidence interval that begins (or ends) precisely at 0. A study that yields a p-value of precisely 0.01 will yield a 99% confidence interval that begins (or ends) precisely at 0. In this sense, reporting an effect size with corresponding confidence intervals can serve as a surrogate for tests of significance (if the confidence interval does not include the nil effect, the study is statistically significant) with the effect size approach focusing attention on the relevant issue. However, by shifting the focus of a report away from significance tests and toward the effect size estimate, we ensure a number of important advantages.

First, effect size focuses attention on the key issue. Usually, researchers and clinicians care about the size of the effect; the issue of whether or not the effect is nil is of relatively minor interest. For example, the clinician might recommend a drug, despite its potential for side effects, if he felt comfortable that it increased the remission rate by some specific amount, such as 20%, 30%, or 40%. Merely knowing that it increased the rate by *some* unspecified amount exceeding 0 is of little import. The effect size with confidence intervals focuses attention on the key index (how large the effect is), while providing likely boundaries for the lower and upper limits of the true effect size in the population.

Second, the focus on effect size, rather than on statistical significance, helps the researcher and the reader to avoid some mistakes that are common in the interpretation of significance tests. Since researchers care primarily about the size of the effect (and not whether the effect is nil), they tend to interpret the results of a significance test as though these results were an indication of effect size. For example, a p-value of 0.001 is assumed to reflect a large effect, while a p-value of 0.05 is assumed to reflect a moderate effect. This is inappropriate because the p-value is a function of sample size as well as effect size. Often, the non-significant p-value is assumed to indicate that the treatment has been proven ineffective. In fact, a non-significant p-value could reflect the fact that the treatment is not effective, but it could just as easily reflect the fact that the study was under-powered.

If power analysis is the appropriate precursor to a study that will test the null hypothesis, then precision analysis is the appropriate precursor to a study that will be used to estimate the size of a treatment effect. This program allows the researcher to take account of both.

Additional Reading

Suggested readings include the following: Borenstein, 1994; Cohen, 1992, 1994; Braitman, 1988; Bristol, 1989; Bulpitt, 1987; Detsky and Sackett, 1985; Feinstein, 1976; Fleiss, 1986a,b; Freiman et. al, 1978; Gardner and Altman, 1986; Gore, 1981; Makuch and Johnson, 1986; McHugh, 1984; Morgan, 1989; Rothman, 1978, 1986a,c; Simon, 1986; Smith and Bates, 1992.

3 The Main Screen

Computing Power Interactively

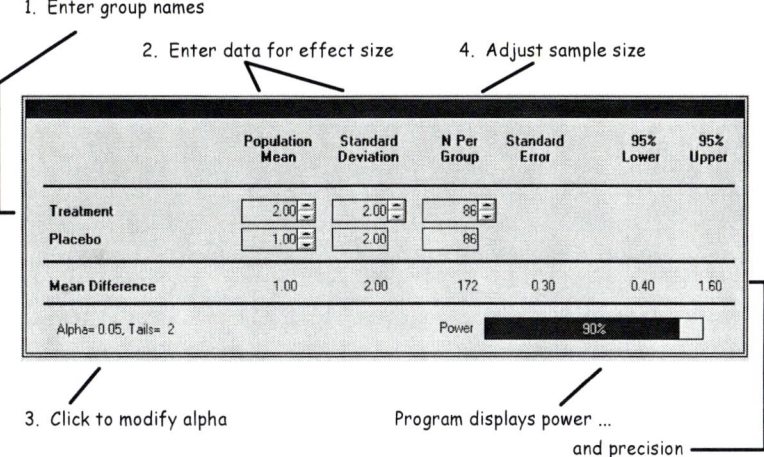

The precise format of the main screen will vary somewhat from one procedure to the next. Following are the key steps for most procedures:

- Optionally, enter names for the group(s).
- Optionally, modify alpha, confidence levels, and tails (choose *Alpha/CI/Tails* from the Options menu or click on the current value).
- Enter data for the effect size.
- Modify the sample size until power and/or precision reach the desired level (or click the *Find N* icon).
- Optionally, save various sets of parameters to the sensitivity analysis list.
- Optionally, click the *Report* icon to create a report.
- Optionally, click the *Make table* or *Make graph* icon to create a table or graph.

Toolbar

Icons on the left side of the toolbar are used to open, save, and print files and copy data to the clipboard. These functions are also available on the File menu.

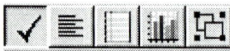

The next set of tools is used to navigate between screens. Click on the check mark to return to the main screen for a procedure. Click on the report, table, or graph tools to create and/or display the corresponding item. These functions are also available on the Tools and View menus.

The next set of icons provides tools for working with the main screen. The first icon is used to find the number of cases required for a given level of power. The second is used to display conventions for small, medium, and large effects. The third opens a list box; the data in the main screen can be saved to this list and later restored. These functions are also available on the Tools menu.

The next set of tools is used to customize the display. The first is used to make text bold, and the second is used to display power using digits rather than a bar graph. These functions are duplicated on the Tools and View menus.

▶   The last set of tools is used to activate various Help functions. Click on these tools to turn on or off the tip of the day, the interactive guide, the interactive summary, and the Help system. These functions are also available on the Help menu.

Interactive Guide

Each procedure includes an interactive guide that can be invoked by clicking the *Wizard* icon or by choosing *Interactive guide* from the Help menu.

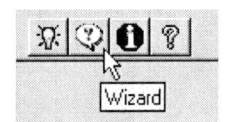

Click the *Wizard* icon to display a series of panels that:
- Introduce the program's features for the current statistical procedure.
- Explain what type of data to enter.
- Provide a running commentary on the planning process.

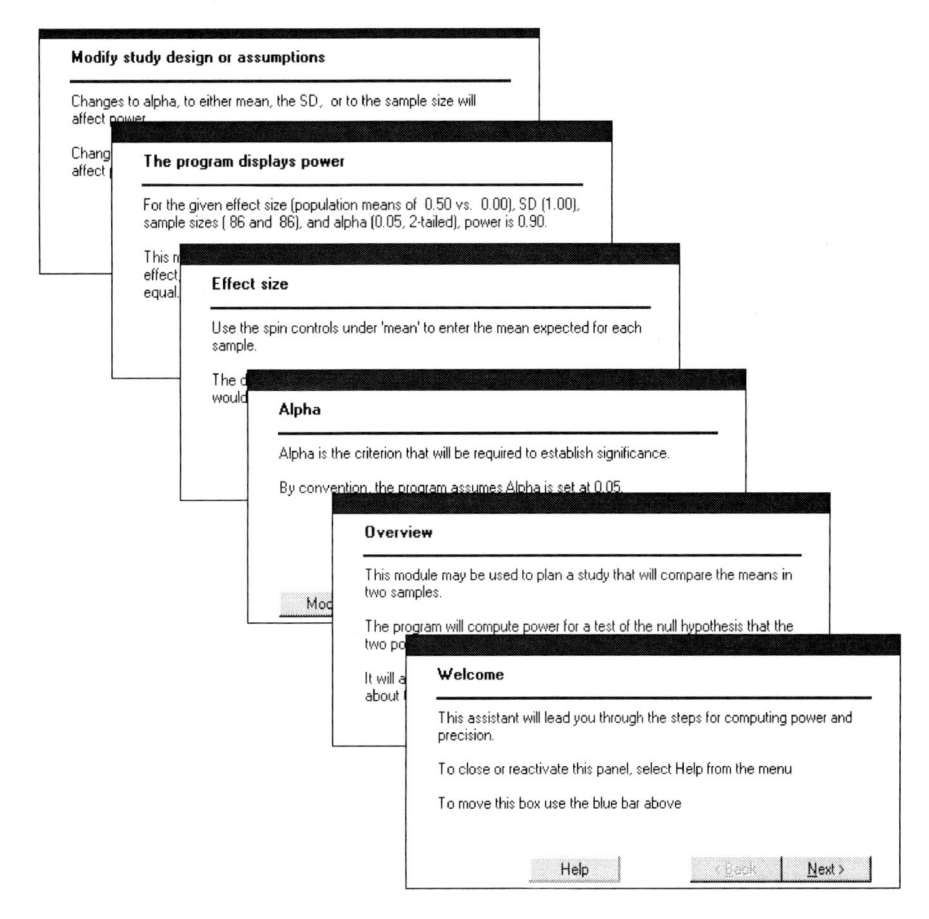

Summary Panel

The program displays a summary panel that offers a text report for the current data. Click the *Display interactive summary* icon or choose *Continuous summary* from the Help menu.

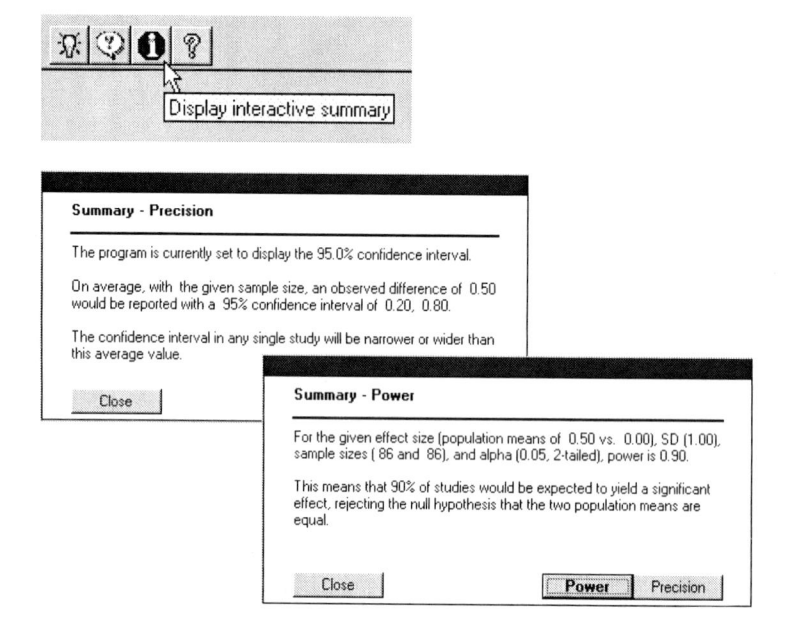

The summary is an on-screen interactive panel. More extensive reports that can be annotated and exported to word processing programs are available in RTF format (see Chapter 6).

Effect Size Conventions

As a rule, the effect size should be based on the user's knowledge of the field and should reflect the smallest effect that would be important to detect in the planned study.

For cases in which the user is not able to specify an effect size in this way, Cohen has suggested specific values that might be used to represent *Small*, *Medium*, and *Large* effects in social science research.

Click the *Show effect size conventions* icon (or choose *Show effect size conventions* from the Tools menu).

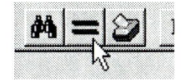

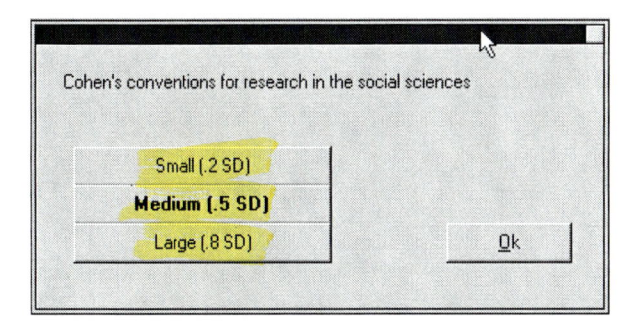

Click *Small*, *Medium*, or *Large* and the program will insert the corresponding effect size into the main screen.

In this program, Cohen's conventions are available for one- and two-sample t-tests, one-sample and two-independent-samples tests of proportions, one- and two-sample correlations, and ANOVA.

Sensitivity Analysis

The user may want to determine how power and precision vary if certain assumptions are modified. For example, how is power affected if we work with an effect (d) of 0.40, 0.45, or 0.50? How is power affected if we work with alpha of 0.05 or 0.01? This process is sometimes referred to as a sensitivity analysis.

The program features a tool to facilitate this process—any set of parameters entered into the main screen can be saved to a list. The list can be used to display power for various sets of parameters—the user can vary alpha, tails, sample size, effect size, etc., with power and precision displayed for each set. The list can be scrolled and sorted. Additionally, any line in the list can be restored to the main screen.

To display the sensitivity analysis list, click the *Show stored scenarios* icon or choose *Display stored scenarios* from the Tools menu.

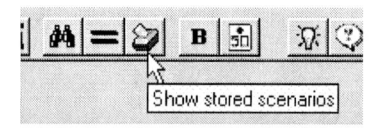

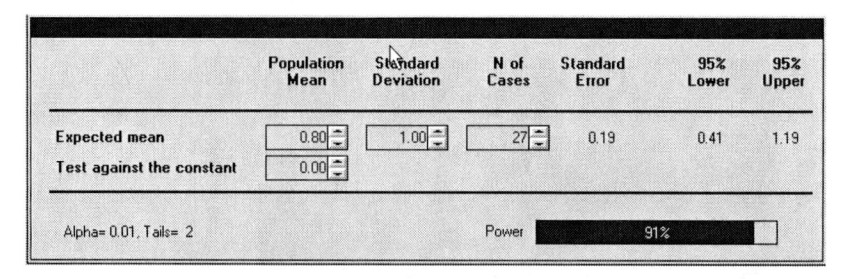

	Population Mean	Standard Deviation	N of Cases	Standard Error	95% Lower	95% Upper
Expected mean	0.80	1.00	27	0.19	0.41	1.19
Test against the constant	0.00					

Alpha= 0.01, Tails= 2   Power 91%

Name	Mean	vs	SD	N Cases	CI Level	Lower	Upper	Tails	Alpha	Power
Small effect	0.20	0.00	1.00	265	.950	0.08	0.32	2	.050	.900
Medium effect	0.50	0.00	1.00	44	.950	0.20	0.80	2	.050	.900
Large effect	0.80	0.00	1.00	19	.950	0.33	1.27	2	.050	.909
Small effect	0.20	0.00	1.00	376	.950	0.10	0.30	2	.010	.901
Medium effect	0.50	0.00	1.00	63	.950	0.25	0.75	2	.010	.901
Large effect	0.80	0.00	1.00	27	.950	0.41	1.19	2	.010	.905

When the list is enabled, the program displays an additional toolbar and a List menu. All functions on the toolbar are also available on the menu.

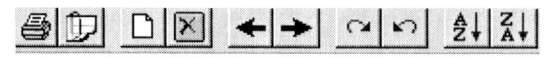

Click the *Print* icon to display the print options, and from the *Forms* tab, select *Stored scenarios* to print the list. Or, click the *Copy list to clipboard* icon to copy the list to the clipboard.

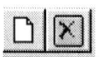

Click the *Clear entire list* icon to clear the entire list. Click the *Erase current line in list* icon to delete the currently selected line.

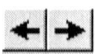

Click the tools above to modify the width of the list columns.

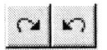

Click the *Store main panel into list* icon to copy the current panel into the next line on the list. This includes all data on effect size, sample size, alpha, confidence intervals, and power. The user can store any number of these lines and then browse them. The adjacent tool will restore the selected line into the main panel, replacing any data currently in the panel.

Note. The program does not store information on computational options, such as the use of z rather than t for t-tests. When a study is restored from the list into the main panel, the options currently in effect will remain in effect.

Click the tools above to select a sort order (descending or ascending). Then, click the title at the top of any column to sort the list by that column.

The sensitivity analysis method is available for one- and two-sample tests of means, correlations, and independent proportions. For other tests (paired proportions, sign test, K × C proportions, ANOVA, and regression), the program does not feature this sensitivity box but does allow the creation of tables using the tables module and allows any scenario to be saved to disk for later retrieval.

Printing Sensitivity Analysis

To print the sensitivity analysis, click the *Print* icon or choose *Print* from the File menu. On the *Forms* tab, select *Stored scenarios*.

Copying to the Clipboard

To copy the sensitivity analysis to the clipboard, click the *Copy list to clipboard* icon on the toolbar. Then switch to a word processing program and press Ctrl-V. Be sure to use the tool on the secondary (sensitivity analysis) toolbar (the clipboard tool on the main toolbar will copy the main screen rather than the list).

Saving Sensitivity Analysis

If the study is saved to disk, the sensitivity analysis is saved as well.

Note. The program can *automatically* generate tables and graphs in which sample size, alpha, and effect size are varied systematically (see Chapter 4 and Chapter 5).

Alpha, Confidence Level, and Tails

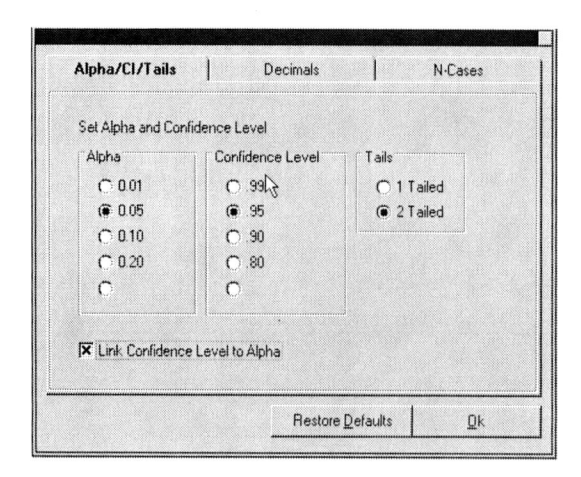

Alpha and Confidence Level

Alpha, confidence intervals, and tails are set from the panel shown above. To activate this panel, click on the value currently shown for alpha, confidence level, or tails on the main screen. Or, choose *Alpha/CI/Tails* from the Options menu.

To set alpha, select one of the standard values or click *Other* and enter a value between 0.001 and 0.499. To set the confidence level, select any of the values offered or click *Other* and enter a value between 0.501 and 0.999.

The check box at bottom of this panel allows you to link the confidence level with alpha. When this option is selected, setting the confidence level at 95% implies that alpha will be set at 0.05; setting the confidence level at 99% implies that alpha will be set at 0.01, and so on. Similarly, any change to alpha is reflected in the confidence level.

Tails

The value set for tails applies to both power and confidence level. Allowed values are 1 and 2. For procedures that are nondirectional (ANOVA, regression, or K × C crosstabs), the tails option is not displayed.

When the test is two-tailed, power is computed for an effect in either direction. When the test is one-tailed, power is computed for a one-tailed test in the direction that will yield higher power.

When the test is two-tailed, the program displays a two-tailed confidence interval (for example, based on a z-value of 1.96 for the 95% interval). When the test is one-tailed, the program displays a one-tailed interval (for example, using a z-value of 1.64). For a one-tailed interval, the program will display both the lower and upper limits, but the user should interpret only one of these. The "interval" extends either from infinity to the upper value shown or from the lower value shown to infinity.

Setting Options for the Number of Cases

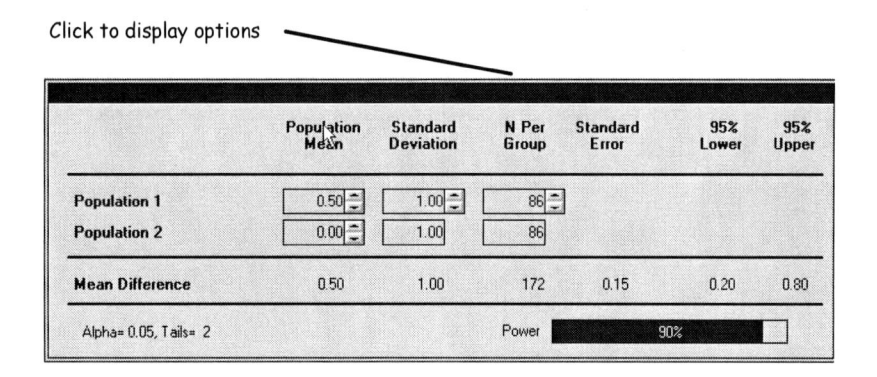

Click to display options

	Population Mean	Standard Deviation	N Per Group	Standard Error	95% Lower	95% Upper
Population 1	0.50	1.00	86			
Population 2	0.00	1.00	86			
Mean Difference	0.50	1.00	172	0.15	0.20	0.80

Alpha= 0.05, Tails= 2   Power   90%

To display the options for setting the sample size, click *N Per Group* or choose *N-Cases* from the Options menu.

Setting the Increment for Spin Control

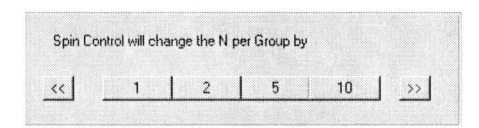

Spin Control will change the N per Group by

<< | 1 | 2 | 5 | 10 | >>

The program allows the user to modify the sample size by using a spin control. The panel shown above allows the user to customize the spin control by setting the size of the increment.

Linking the Number of Cases in Two Groups

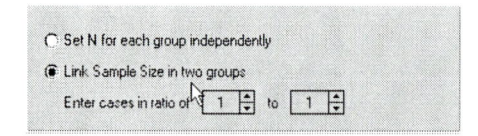

Initially, the program assumes that cases will be allocated to two groups in a ratio of 1:1. This panel allows the user to set a different ratio (for example, 1:2 or 3:5). (Note that power is based on the harmonic mean of the two sample sizes, which means that it is controlled primarily by the lower of the two sample sizes.) As long as the sample sizes are linked, the program will expect (and accept) a sample size for the first group only.

The user can also set the number of cases for each group independently of the other.

Note. Some of the program's features (Find N, Table, and Graph) will not operate unless the sample sizes in the two groups are linked, but these features do not require that cases be assigned to the two groups in even numbers.

Finding the Number of Cases Automatically

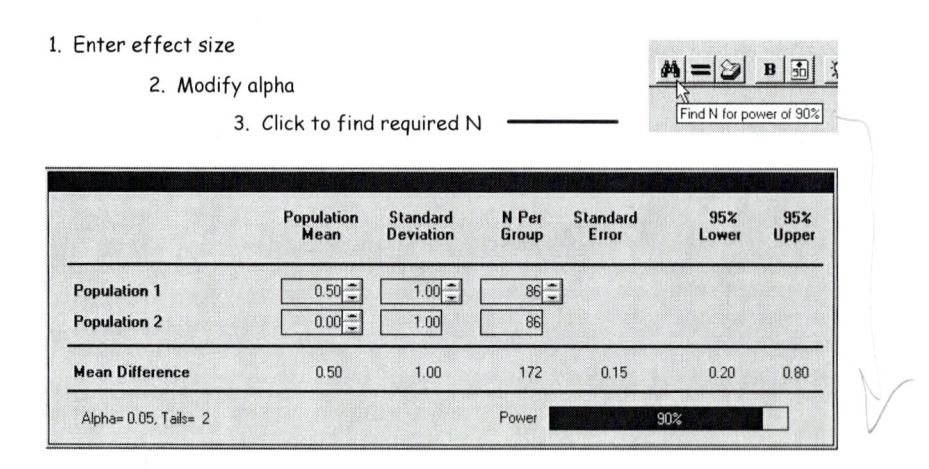

1. Enter effect size
2. Modify alpha
3. Click to find required N

The program will find the number of cases required for the default value of power. Enter an effect size, modify alpha, and click the *Find N* icon. By default, the program assumes that the user wants power of 80%, but this can be modified (see "Modifying the Default Value of Power" below). Note that the number of cases required for precision is usually different from the number of cases required for power. The number of cases required for precision can be found by using the spin control to modify the number until precision is appropriate.

In situations in which the user has specified that cases should be allocated unevenly to the two groups (for example, in a ratio of 1:2), the program will honor this allocation scheme in finding the required sample size.

Modifying the Default Value of Power

To temporarily modify the default value for required power, press Ctrl-F or choose *Find N* from the Tools menu.

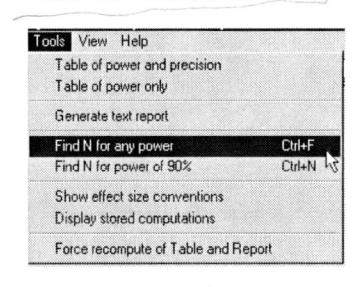

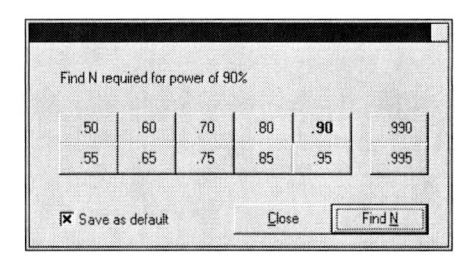

- To find the number of case required for a specific power, click the power. For example, to find the number of cases for power of 95%, click 0.95. The required number is shown immediately.
- To temporarily change the required power to 95%, select *Save as default* and then click 0.95. This new power will remain in effect as long as you remain within the module. (The *Find N* icon and the *Find N* menu option will read "95%" as well.) Power will revert to the default when you leave the module.
- To set 95% as the default, choose *Preferences* from the Options menu. This value will take effect with the next module and will be used as the default value for future sessions.

Printing the Main Screen

To print the data on the main screen, click the *Print* icon or choose *Print* from the File menu.

Copying to the Clipboard

To copy the data from the main screen to the clipboard, click the *Copy to clipboard* icon or choose *Clipboard* from the File menu. Then switch to a word processing program and choose *Paste* from the Edit menu. The data from the main screen cannot be pasted into the program's report screen.

Saving and Restoring Files

Any file can be saved to disk and later retrieved. Click the *Save file* icon or choose *Save study data* from the File menu.

The program saves the basic data (effect size, alpha, sample size, and confidence level), as well as any sets of data that had been stored, to the sensitivity analysis box.

Note. The program does not store computation options; the options in effect when the file is later retrieved will be used to recompute power at that time.

The *Save study data* menu command does not save tables, graphs, or reports. These can be saved in their own formats. They can be:

- Copied to a word processing program and saved with the word processing document
- Recreated at a later point by retrieving the basic data and clicking the icon for a report, table, or graph

The file must be saved from within a module (for example, from within the t-test module or the ANOVA module). Files can be retrieved from any point in the program, including the opening screen.

Data files are stored with an extension of *.pow*. Double-clicking on one of these files in the Windows Explorer will cause Windows to launch *Power and Precision* and load the data file.

4 Tables

The program can create tables and graphs with minimal input from the user. To create a table using the program's default settings:

- Enter an effect size.
- Optionally, modify alpha, tails, or the confidence level.
- Click the *Make table* icon.

Figure 4.1 Creating a table using default settings

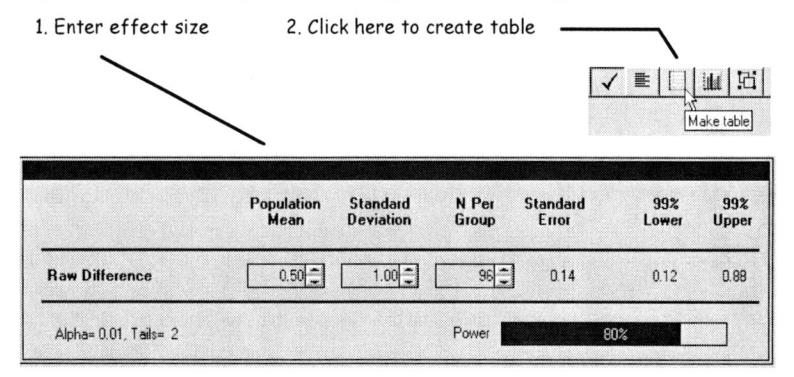

To create a table, the program works with the effect size and alpha specified by the user on the main screen. It uses these values to set a range of values (start, increment, and ending value) for the sample size that will show power beginning at a low value and increasing to approximately 99%.

The program allows the user to customize the table in various ways. The first step in this process (selecting either of two basic styles) may be taken before or after the default table is created. All other modifications are made after the default table is displayed.

Table Style

The program features two basic table styles: tables of power and precision as a function of sample size, and tables of power only. When both table styles are available for a procedure, you can select the style you want. To select a style before creating a table, choose *Table of power and precision* or *Table of power only* from the Tools menu. To select a style after the default table has been created, choose *Style* from the Modify menu.

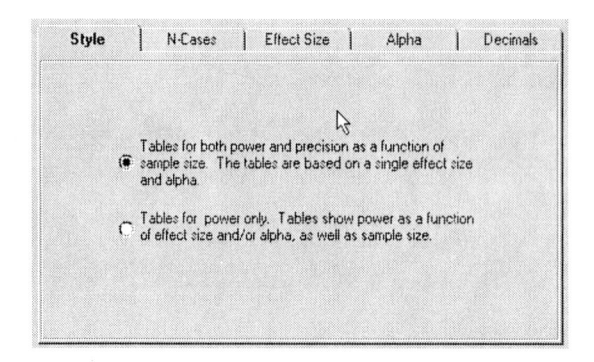

Power and Precision

In this style, the user is limited to a single value for the effect size and alpha and/or tails. For this value, the program will generate a column of power and also a set of columns showing the precision (standard error and confidence intervals) for each effect size. The standard error is displayed only when it is meaningful (that is, it is displayed for the log odds ratio but not for the odds ratio). An example is shown in Figure 4.2.

Power Only

This style allows the user to specify one to five values for the effect size and also one to four values for alpha and/or tails. The table will automatically generate all possible combinations of these values. After producing this kind of table, the program can generate a graph showing power as a function of effect size and the number of cases (using the first value of alpha) or as a function of alpha and the number of cases (using the initial effect size). An example is shown in Figure 4.3.

Tables   33

Figure 4.2   Power and precision

Select this style

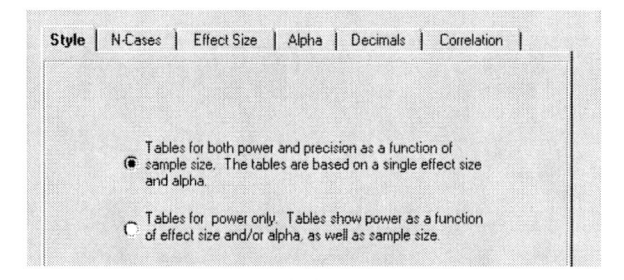

To yield this table

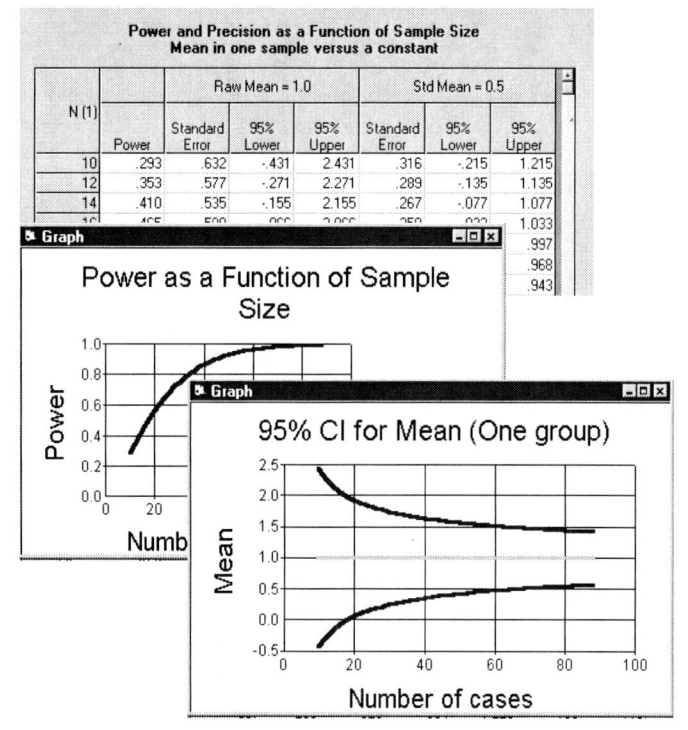

And these graphs

34   Chapter 4

Figure 4.3   Power only

Select this style

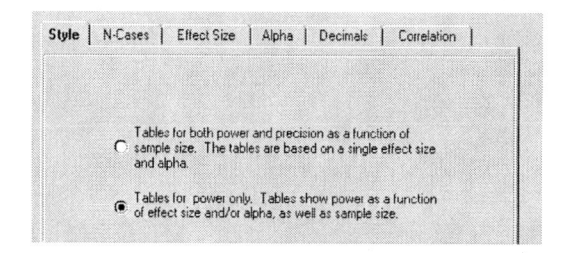

Tables for both power and precision as a function of sample size. The tables are based on a single effect size and alpha.

● Tables for power only. Tables show power as a function of effect size and/or alpha, as well as sample size.

To yield this table

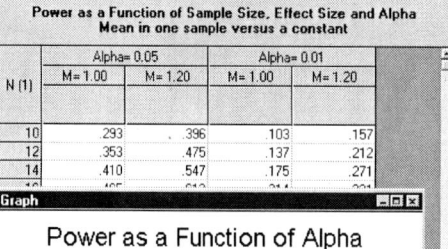

Power as a Function of Sample Size, Effect Size and Alpha
Mean in one sample versus a constant

N (1)	Alpha= 0.05 M=1.00	Alpha= 0.05 M=1.20	Alpha= 0.01 M=1.00	Alpha= 0.01 M=1.20
10	.293	.396	.103	.157
12	.353	.475	.137	.212
14	.410	.547	.175	.271

And these graphs

Power as a Function of Alpha and N

Power as a Function of Effect Size and N

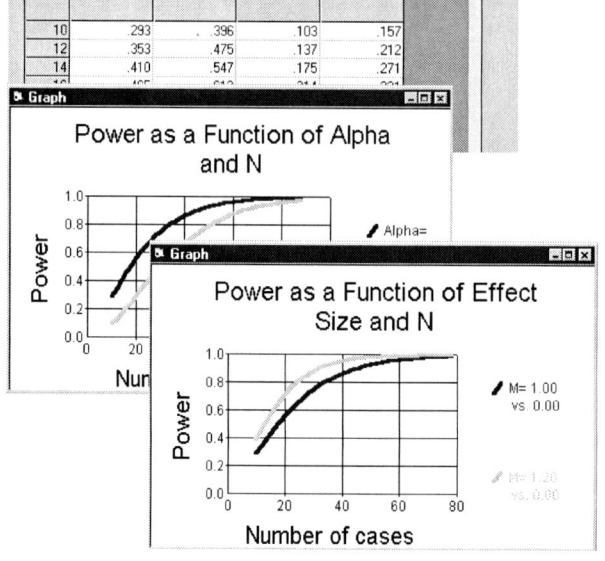

Effect Size

Enter one to five sets of values for the effect size. In the example below, the mean for group 1 is either 1.0 (columns 1 and 2) or 1.2 (columns 3 and 4). The mean for group 2 is constant at 0.5. The standard deviation is either 2 (columns 1 and 2) or 1.5 (columns 3 and 4). The result of entering multiple effect sizes is shown below.

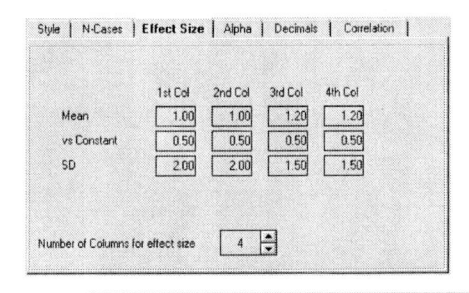

To display power for only one value of effect size, reset the spin control to 1.

Note. To display power as a function of multiple effect size values, alpha, or tails, the table style must be set to power only. When the table style is set to power and precision, only the first combination of effect size, alpha, and tails is used.

The sequence for effect size should be highest to lowest. This ensures that higher values of power will precede lower values, and will yield graphs in which the sequence of lines corresponds to the sequence shown in the legend.

Alpha and/or Tails

Enter one to four sets of values for alpha and/or tails. In this example, alpha is either 0.05 (columns 1 and 2) or 0.01 (columns 3 and 4). For either value of alpha, the user has specified both one and two tails. The resulting table is shown below.

To display power for only one value of alpha, reset the spin control to 1.

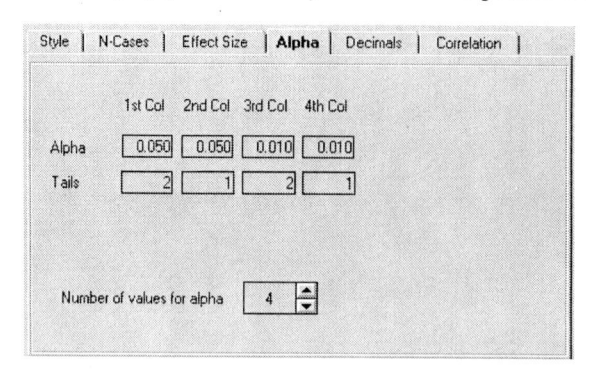

Note. To display power as a function of multiple effect size values, alpha, or tails, the table style must be set to power only. When the table style is set to power and precision, only the first combination of effect size, alpha, and tails is used.

When entering more than one value for alpha, the sequence should be from highest to lowest (for example, 0.20, 0.10, 0.05, ...). Enter the one-tailed values before the two-tailed values. This ensures that higher values of power will precede lower values and will yield graphs in which the sequence of lines corresponds to the sequence shown in the legend.

Varying Effect Size and Alpha

When more than one value has been specified for both effect size and alpha, the program generates a table with all possible combinations of the two. Here, for example, the user specified two values for the mean (1.0 and 1.2) and two values for alpha (0.05 and 0.01). The program generates four columns: alpha of 0.05 for either mean and alpha of 0.01 for either mean.

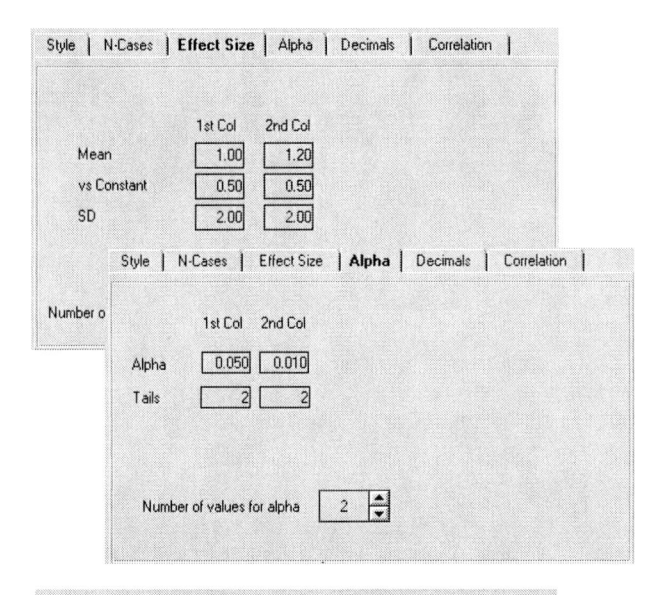

| Style | N-Cases | **Effect Size** | Alpha | Decimals | Correlation |

	1st Col	2nd Col
Mean	1.00	1.20
vs Constant	0.50	0.50
SD	2.00	2.00

| Style | N-Cases | Effect Size | **Alpha** | Decimals | Correlation |

	1st Col	2nd Col
Alpha	0.050	0.010
Tails	2	2

Number of values for alpha 2

Power as a Function of Sample Size, Effect Size and Alpha
Mean in one sample versus a constant

N (1)	Alpha= 0.05 M= 1.00	Alpha= 0.05 M= 1.20	Alpha= 0.01 M= 1.00	Alpha= 0.01 M= 1.20
10	.109	.168	.028	.049
20	.186	.318	.061	.126
30	.263	.458	.100	.222
40	.338	.579	.144	.324
50	.410	.679	.193	.427
60	.478	.760	.244	.524
70	.541	.823	.297	.611

The graphs, however, display only one of these factors at a time. In the example above, the user can produce a graph showing power as a function of effect size and for the initial value of alpha (0.05). Or, the user can produce a graph showing power as a function of effect size and alpha and/or tails.

Sample Size

The program automatically sets the range to be displayed for the sample size (in this example, 10 by 10 to 250). The range selected will produce an appropriate graph for power, taking into account both the range of effect sizes and the range of values for alpha. The same range is then applied to graphs of confidence intervals.

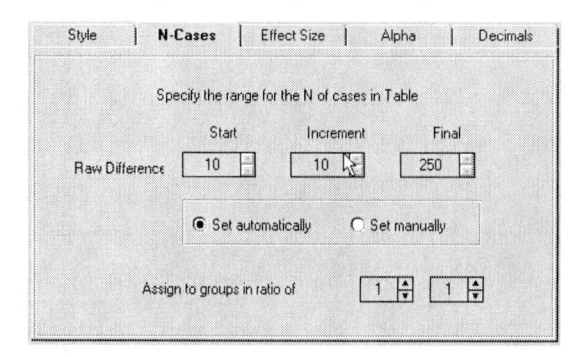

The user has the option of setting the range manually by using the panel shown above. Select *Set manually* to activate the panel. At any time, select *Set automatically* to revert to automatic selection.

Note. For two-sample tests, the program will allow the user to specify the ratio of cases entered into the two groups. The range shown applies to the first group. The automatic range function will work with any ratio set by the user.

Tip

You may want to set the range manually when focusing on one portion of the range just before printing the table. Otherwise, it is recommended that you set the range of values for the number of cases automatically so that it can adjust to any changes in the effect size or alpha.

Computational Options

Computational options are set from the main screen (choose *Computational formulas* from the Options menu). Some computational options, such as Fisher's exact test for 2×2 proportions (those requiring substantial time for iterations), are not available for tables.

Modifying Titles

Before printing, the user can modify the title set by the program. Choose *Table Title* from the Modify menu.

Printing

Click the *Print* icon, or choose *Print* from the File menu. On the *Forms* tab, select *Table*. Using the toolbar, you can modify the width of the columns before printing.

Copying to the Clipboard

Click the *Copy to clipboard* icon, or choose *Clipboard* from the File menu to copy a table to the clipboard. Then switch to a word processing program and paste the table into that program.

The table is copied to the clipboard as an object rather than as text. While it can be copied into any Windows-based word processing program that supports objects (for example, Microsoft Word or WordPerfect), it cannot be copied into a report within this program (in release 1.0). Within the word processing program, you should be able to modify the physical size of the table, although the word processing program may require that it be confined to a single page.

Saving to a File

To save a table to a file, copy it to the clipboard and then into a word processing program.

Generating a Graph

After the table has been created, click on the Graph menu to display a list of available graphs.

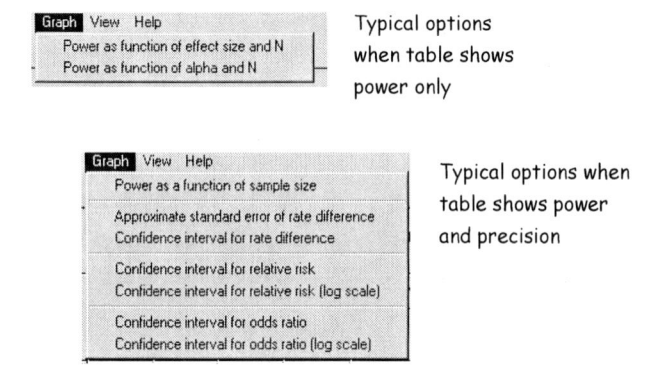

Typical options when table shows power only

Typical options when table shows power and precision

The graphs will mirror the current table. If the table shows power for multiple effect sizes, the options will include *Power as a function of effect size and N*. If the table shows confidence intervals, the selections might include *Standard error of the mean* and *Confidence interval for the mean*.

5 Graphs

The program can create graphs with minimal input from the user. To create a graph using the program's default settings:

- Enter an effect size.
- Optionally, modify alpha, tails, or the confidence level.
- Click the *Make graph* icon.

Every graph is based on a table. When you create a graph using this shortcut, the program will create the default table and then proceed automatically to create the corresponding (default) graph.

Figure 5.1 Creating a graph using default settings

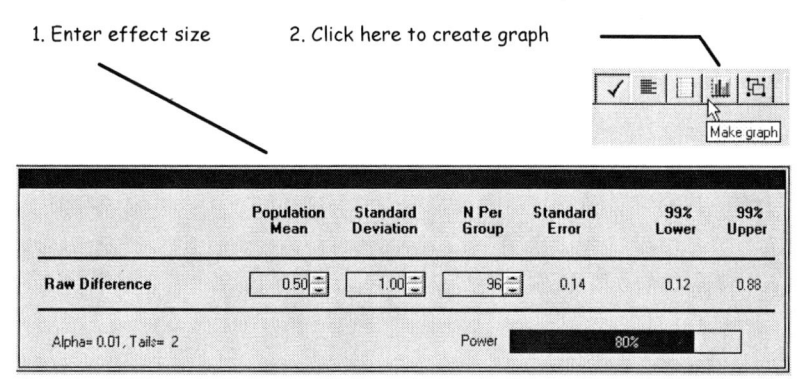

Chapter 5

The program can create various types of tables (see Chapter 4) and corresponding graphs. The first menu in Figure 5.2 shows the kinds of graphs available when the table style is set to power *only*. The second menu shows the kinds of graphs available when the table style is set to power *and* precision.

Figure 5.2 Types of graphs available

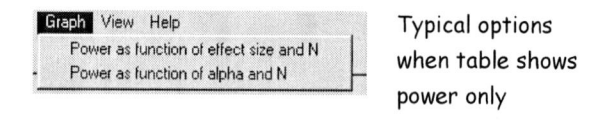

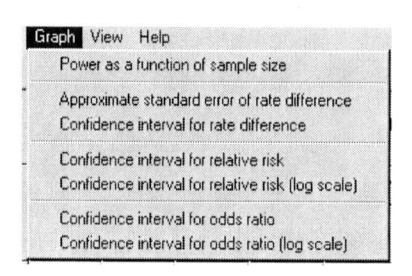

Typical options when table shows power only

- Power as function of effect size and N
- Power as function of alpha and N

Typical options when table shows power and precision

- Power as a function of sample size
- Approximate standard error of rate difference
- Confidence interval for rate difference
- Confidence interval for relative risk
- Confidence interval for relative risk (log scale)
- Confidence interval for odds ratio
- Confidence interval for odds ratio (log scale)

Figure 5.3 Power as a function of sample size

Power as a Function of Sample Size
Two sample t-test

Number of cases per group
Mean Diff=0.5 SD=1.0 Alpha=.05 Tails=2

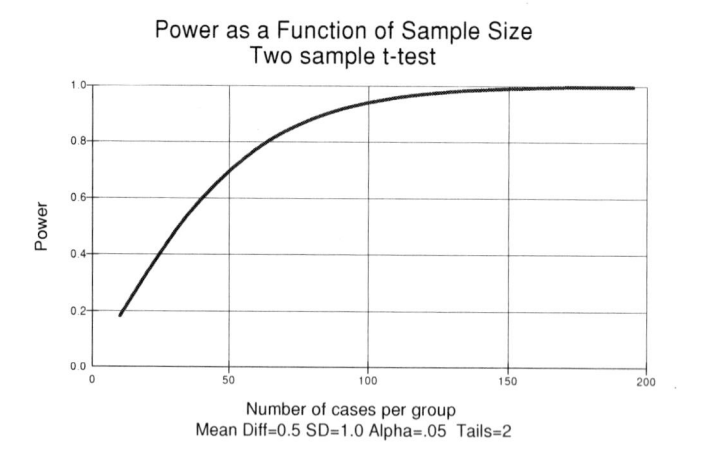

Figure 5.3 is an example of a default graph. It is available with either table style.

Figure 5.4 Power as a function of effect size and sample size

Power as a Function of Effect Size and N
Two sample t-test

(Diff= 0.50, Diff= 0.40, Diff= 0.30)

Number of cases per group
SD=1.0 Alpha=.05 Tails=2

To create the graph shown in Figure 5.4, set the table style to power only and specify multiple values for effect size. If multiple values are set for both effect size and alpha, the graph will show power as a function of effect size for the initial value of alpha.

Figure 5.5 Power as a function of alpha and sample size

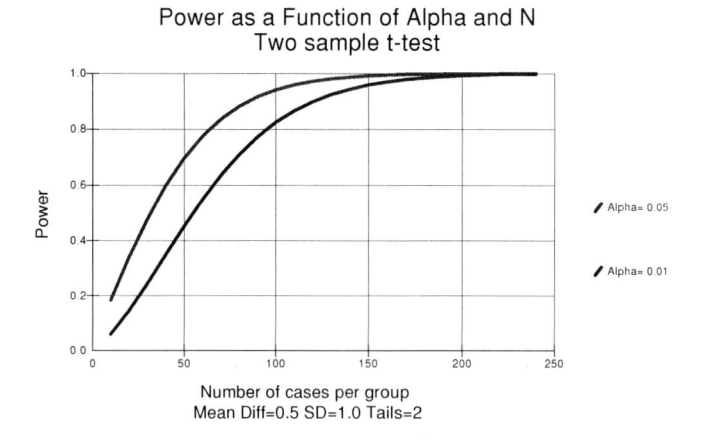

Power as a Function of Alpha and N
Two sample t-test

(Alpha= 0.05, Alpha= 0.01)

Number of cases per group
Mean Diff=0.5 SD=1.0 Tails=2

To create the graph shown in Figure 5.5, set the table style to power only and specify multiple values for alpha and/or tails. If multiple values are set for both effect size and alpha, the graph will show power as a function of effect size for the initial value of alpha.

Figure 5.6 Standard error as a function of sample size

Standard Error of Mean Difference — Two sample t-test

(Number of cases per group; SD Within Groups =1.0)

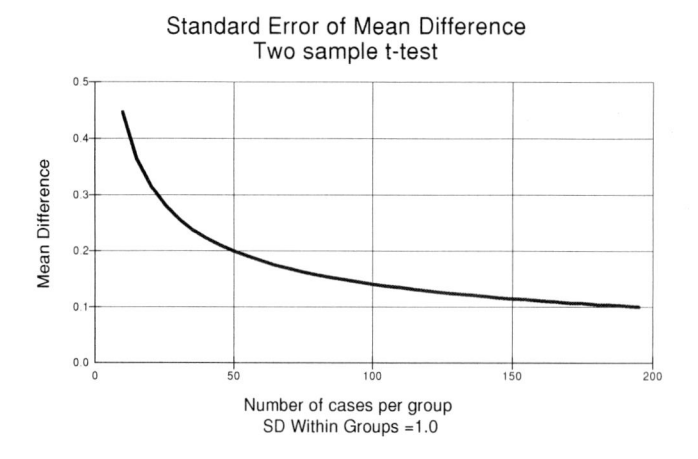

To create the graph shown in Figure 5.6, set the table style to power and precision. Then choose *Standard Error* from the Graph menu.

Figure 5.7 Precision as a function of sample size

Expected 95% Confidence Interval for Raw Difference — Two sample t-test

(Number of cases per group; Mean Diff=0.5 SD=1.0 Alpha=.05 Tails=2)

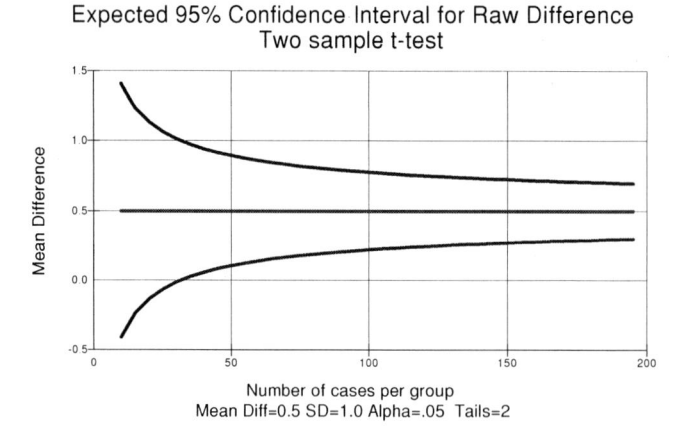

To create the graph shown in Figure 5.7, set the table style to power *and* precision. Then choose *Confidence Intervals* from the Graph menu. The confidence level (95% in this example) is initially taken from the main screen. It can be modified by choosing *Alpha* from the Modify menu.

The precision shown is the expected precision for the given sample size. The actual precision will vary from study to study (see "Tolerance Intervals" on p. 16 in Chapter 2.)

Customizing Graphs

To customize a graph, first create and modify a table. The program then allows you to create a graph based on the modified table.

The user can set the table style (see Chapter 4).

When the table includes data for both power and precision, the user can create graphs that show power, the standard error of the effect, and the expected precision of the effect as a function of sample size.

When the table includes data for power only, the user can create graphs that show power as a function of effect size and sample size or power as a function of alpha and sample size.

Modifying Content

The Modify menu allows the user to modify the content of the underlying table as follows:

- Set the style of the graphs. (One style allows graphs to show either power or precision. The other style allows graphs to show power only as a function of effect size and alpha).
- Set the effect size. When entering more than one effect size, enter values from largest to smallest so that the legend sequence will correspond to the sequence of power lines.
- Set values for alpha and/or tails. When entering more than one value for alpha, enter them in the sequence 0.10, 0.05, and 0.01. If applicable, enter one-tailed values before two-tailed. This will force the legend sequence to correspond to the sequence of power lines.
- Set the range of cases. The program will set the range automatically unless the user chooses *Set manually*.

Formatting Titles, Legends, and Colors

The Format menu is used to modify graph titles, legends, and colors. First, create the default graph. Then, choose the option you want from the Format menu.

Figure 5.8   Modifying titles

	Titles	Legend	Color
Top	Power as a Function of Sample Size		☒
	Two sample t-test		☒
Left	Power		☒
Bottom	Number of cases per group		☒
	Mean1=0.2 Mean2=0.0 SD=1.0 Alpha=.05 Tails=2		☒
Right	Mean1=0.2 Mean2=0.0 SD=1.0 Alpha=.05 Tails=2		☐

The user can edit any of the program's default titles. The user can also determine whether or not a particular title will print at all (select or deselect the corresponding check box).

Notice that one of the default titles is assigned to two locations (bottom and right). By default, this title prints at the bottom, but by deselecting one check box and selecting another, the user can choose to have this title print in another location.

Figure 5.9   Modifying legends

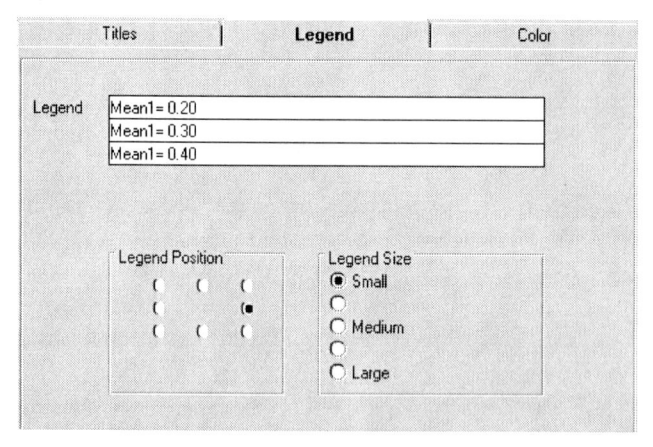

	Titles	Legend	Color
Legend	Mean1= 0.20		
	Mean1= 0.30		
	Mean1= 0.40		

Legend Position

Legend Size
- ● Small
- ○ Medium
- ○ Large

This panel allows the user to edit the default legends or to modify the size and placement of the legend text.

Figure 5.10 Modifying colors

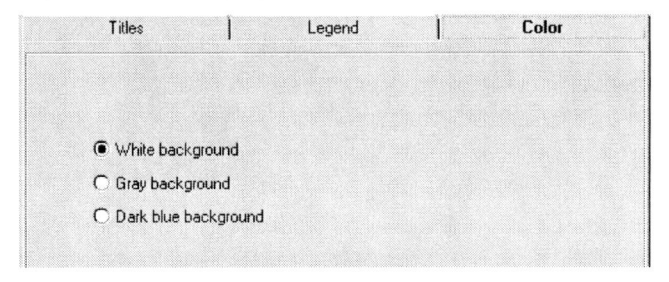

[Dialog box showing tabs: Titles | Legend | Color, with radio button options: ● White background, ○ Gray background, ○ Dark blue background]

The graph is always displayed in color. The program includes an option for monochrome printing, which is selected from the Print menu.

Printing Graphs

Click the *Print* icon or choose *Print* from the File menu.

The Print panel includes an option to print a graph in monochrome even if it is being displayed in color. In this case, the program will add symbols to distinguish between different lines. This option should be selected for graphs showing power as a function of effect size and alpha because the symbols are needed to distinguish between lines. For graphs showing confidence intervals, you may want to print in color (even if the actual print will be in monochrome) because the symbols would be superfluous.

Copying Graphs to the Clipboard

Click the *Copy to clipboard* icon or choose *Clipboard* from the File menu.

Then switch to a word processing program and press Ctrl-V (or use the program's Edit menu) to paste the graph directly into that application.

Saving Graphs to Disk

Choose *Save graph to disk* from the File menu. The file will be saved in Windows metafile (*.wmf*) format.

Do not save the graph by using the *Save file* icon or the *Save study data* menu item. These save the study parameters rather than the graph.

Note. The graph shows both the lower and upper levels for a confidence interval. When the confidence level is one-tailed, a lower boundary and an upper boundary are shown, but only one is meaningful—the interval extends from the value shown on one side to infinity on the other.

6 Reports

The program can generate a report that summarizes the results of the power and precision analysis in a format that is compatible with most Windows word processing programs.

Figure 6.1 Creating a report

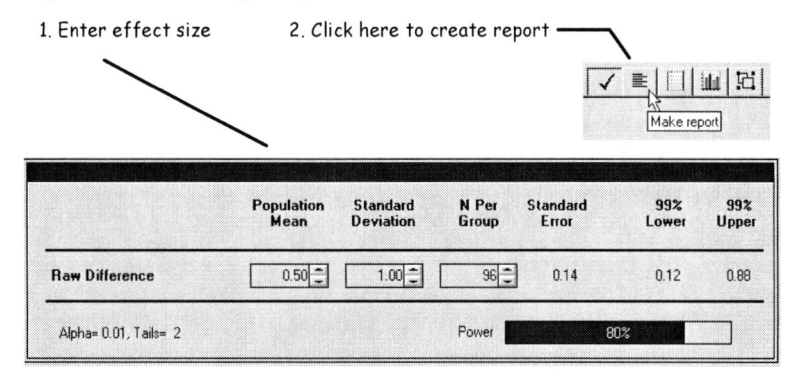

- On the main screen, enter an effect size.
- Optionally, modify the group names, alpha, tails, confidence level, and sample size.
- Modify the study parameters until power and/or precision are acceptable.
- Click the *Make report* icon on the toolbar, or from the Tools menu choose *Generate text report*.
- You can edit the report using standard word processing procedures.
- The report is created in rich text format (RTF), which is compatible with most word processing programs.

If you generate additional reports, the new reports will be added to the end of the file. Figure 6.2 shows sample reports.

Power for a test of the null hypothesis

One goal of the proposed study is to test the null hypothesis that the two population means are equal. The criterion for significance (alpha) has been set at 0.01. The test is 2-tailed, which means that an effect in either direction will be interpreted.

With the proposed sample size of 96 and 96 for the two groups, the study will have power of 80.5% to yield a statistically significant result.

This computation assumes that the mean difference is 0.5 and the common within-group standard deviation is 1.00.

This effect was selected as the smallest effect that would be important to detect, in the sense that any smaller effect would not be of clinical or substantive significance. It is also assumed that this effect size is reasonable, in the sense that an effect of this magnitude could be anticipated in this field of research.

Precision for estimating the effect size

A second goal of this study is to estimate the mean difference between the two populations. On average, a study of this design would enable us to report the mean difference with a precision (99.0% confidence level) of plus/minus 0.37 points.

For example, an observed difference of 0.5 would be reported with a 99.0% confidence interval of 0.13 to 0.87.

The precision estimated here is the median precision. Precision will vary as a function of the observed standard deviation (as well as sample size), and in any single study will be narrower or wider than this estimate.

Figure 6.2   Sample reports

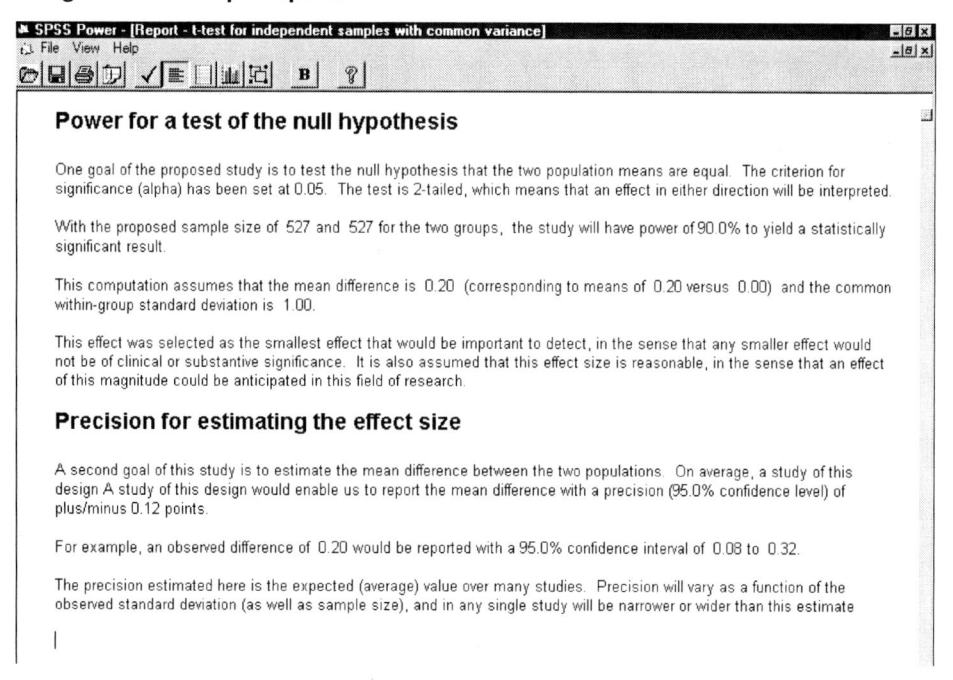

> **Power for a test of the null hypothesis**
>
> One goal of the proposed study is to test the null hypothesis that the two population means are equal. The criterion for significance (alpha) has been set at 0.05. The test is 2-tailed, which means that an effect in either direction will be interpreted.
>
> With the proposed sample size of 527 and 527 for the two groups, the study will have power of 90.0% to yield a statistically significant result.
>
> This computation assumes that the mean difference is 0.20 (corresponding to means of 0.20 versus 0.00) and the common within-group standard deviation is 1.00.
>
> This effect was selected as the smallest effect that would be important to detect, in the sense that any smaller effect would not be of clinical or substantive significance. It is also assumed that this effect size is reasonable, in the sense that an effect of this magnitude could be anticipated in this field of research.
>
> **Precision for estimating the effect size**
>
> A second goal of this study is to estimate the mean difference between the two populations. On average, a study of this design A study of this design would enable us to report the mean difference with a precision (95.0% confidence level) of plus/minus 0.12 points.
>
> For example, an observed difference of 0.20 would be reported with a 95.0% confidence interval of 0.08 to 0.32.
>
> The precision estimated here is the expected (average) value over many studies. Precision will vary as a function of the observed standard deviation (as well as sample size), and in any single study will be narrower or wider than this estimate

Later chapters in this manual include sample reports generated for each statistical procedure.

Tip

The program allows the user to incorporate citations to all analyses referenced in the report (choose *Add references* from the Edit menu).

Printing Reports

Click the *Print* icon, or from the File menu choose *Print*. On the *Forms* tab, select *Report* (this option will be enabled only if a report has been created).

Copying Reports to the Clipboard

Highlight all or part of the report (if none of the report is selected, the entire report will be copied). Then click the *Copy to clipboard* icon, or from the File menu choose *Clipboard*. Switch to a word processing program and paste the report into a file.

Saving Reports

Choose *Save report to disk* from the File menu. The report will be saved in RTF format and can later be imported into a word processing program.

Tip

Do not save to disk using the *Save file* icon or the *Save study data* option on the File menu. These functions save the study parameters rather than the report.

7 T-Test for One Group

Selecting the Procedure

To display the available procedures, choose *New analysis* from the File menu.

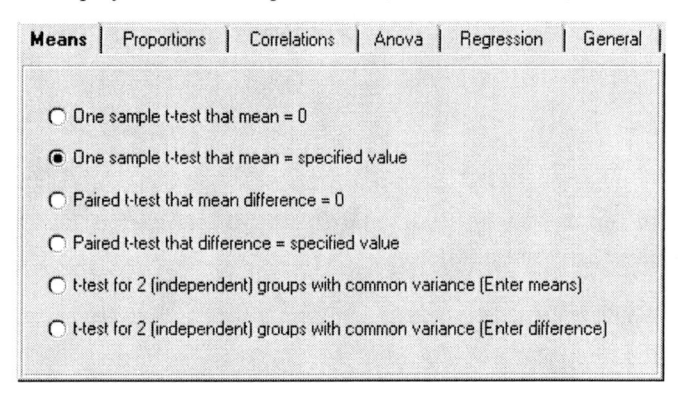

| Means | Proportions | Correlations | Anova | Regression | General |

- ○ One sample t-test that mean = 0
- ◉ One sample t-test that mean = specified value
- ○ Paired t-test that mean difference = 0
- ○ Paired t-test that difference = specified value
- ○ t-test for 2 (independent) groups with common variance (Enter means)
- ○ t-test for 2 (independent) groups with common variance (Enter difference)

Application

Figure 7.1 One sample t-test

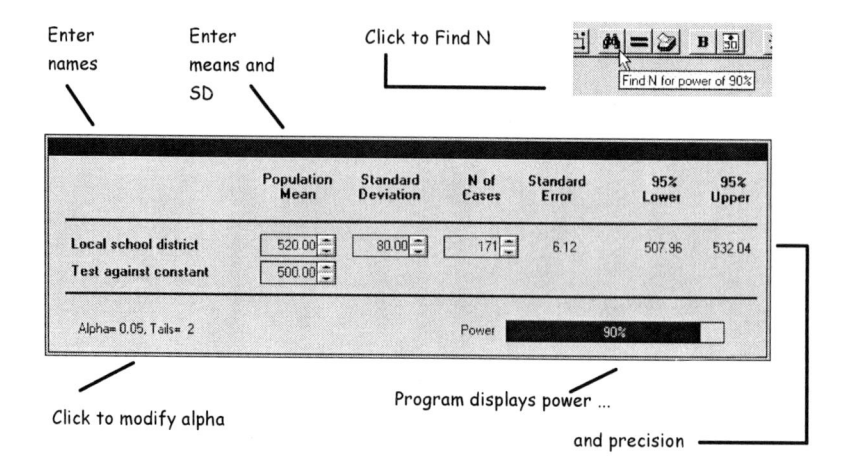

Enter names

Enter means and SD

Click to Find N

Find N for power of 90%

	Population Mean	Standard Deviation	N of Cases	Standard Error	95% Lower	95% Upper
Local school district	520.00	80.00	171	6.12	507.96	532.04
Test against constant	500.00					

Alpha= 0.05, Tails= 2   Power   90%

Click to modify alpha

Program displays power ...

and precision

The program includes two versions of the one-group t-test. The first assumes that the mean will be tested against the nil value (0). The second allows the user to test the mean against any specific (null) value.

Effect Size

The effect size for t-tests is the standard difference, d, defined as the mean difference divided by the standard deviation.

In theory, the effect size index extends from 0 (indicating no effect) to infinity. In practice, of course, d is limited to a substantially smaller range, as reflected in the conventions suggested by Cohen for research in the social sciences—small (0.20), medium (0.50), and large (0.80).

The program requires the user to enter the mean and standard deviation (SD) for the population to be tested. If the mean will be tested against a mean of 0, no additional data are required for the effect size. If the mean will be tested against a specific (nonzero) value, the researcher must supply this value as well.

Tip

To enter the standard difference (d) directly, set the program to enter the raw difference and provide a value of 1.0 for the standard deviation. In this case, the raw difference is equal to d.

Alpha, Confidence Level, and Tails

Click *Alpha*, *Tails*, or the confidence level to modify these values. The value set for tails applies to both the power analysis and the confidence interval.

Sample Size

The spin control adjacent to the N of Cases value can be used to modify the sample size quickly. The size of the increment can be specified by the user (click *N of Cases*).

Click the *Find N* icon to have the program find the number of cases required for the default level of power. The default value for power is 80%, but this can be modified temporarily (Ctrl-F) or permanently (choose *Preferences* from the Options menu).

Tolerance Interval

The confidence interval displayed by the program is the median interval and is shown balanced about the expected mean. For example, assume that the user enters an expected mean of 100, with SD = 10 and N = 40. The program displays the 95% confidence interval as 96.83 to 103.17 (which is equal to the mean plus or minus 3.17).

Choose *Tolerance intervals* from the View menu. The program shows that the 50% tolerance interval for the 95% confidence interval is 6.34 (corresponding to the distance from 96.83 to 103.17), which means that in 50% of the studies, the 95% confidence interval will be no wider than 6.34 points.

Additionally, it shows that the 80% tolerance interval is 6.96, which means that in 80% of all studies, the 95% confidence interval will be no wider than 6.96.

To modify the confidence level, click on the value shown, or choose *Alpha/Tails* from the Options menu. To modify the tolerance level, change the value shown in the tolerance interval box.

Because a one-tailed "interval" actually extends from minus infinity to some value above the mean, or from some value below the mean to plus infinity, it is not useful here. We could, however, consider the distance from the observed mean to the single boundary. If the user selects a one-tailed confidence interval, the median distance from the observed mean to the single boundary will be no greater than 2.64 in 50% of samples and no greater than 2.90 in 80% of samples.

As the sample size increases, not only are we able to estimate the mean more precisely (which yields a narrower confidence interval), but the dispersion of the sample standard deviation about the expected standard deviation decreases; this means that the width of the confidence interval shows less variation from sample to sample. For example, with N = 200, the 50% and 80% tolerance intervals are quite similar, at 2.78 and 2.90, respectively (assuming that SD = 10,0 and using two tails).

When the user specifies that the variance is known (choose *Computational formulas* from the Options menu), the confidence interval is computed using the population variance specified by the user. In this case, the confidence interval will not vary from sample to sample, and the 80% (or any other) tolerance interval is identical to the 50% tolerance interval.

Computational Options for Power

By default, the program assumes that the population variance is unknown (will be estimated based on the observed variance), which is normally the case. The program allows the user to specify that the population variance is known (technically, this would be a z-test rather than a t-test). In this case, the program will work with the z-distribution rather than the t-distribution for computing power and confidence intervals. In practice, this distinction will have a substantial impact only if the number of cases is quite small (under 20–30) and the effect size is large. With larger sample sizes, the impact on computations may not be evident.

To switch between the t-test and the z-test, choose *Computational formulas* from the Options menu.

This module applies exclusively to the t-test for one group. The formulas used are identical to those used by the paired t-test module, but the reports and tools are optimized for each module. The present module should *not* be used for the two-group t-test because the computational formulas are different.

Computational Options for Precision

The computational options for precision are identical to those for power. By default, the program works with the t-distribution (for estimated variance), but the user can choose to work with the z-distribution (for known variance). The option set for power is applied to precision as well.

Options for Screen Display

The program will display the standard difference, d. Choose *Customize screen* from the Options menu.

Example

A school administrator wants to know whether students in his school district are scoring better or worse than the national norm of 500 on the SAT. He decides that a difference of 20 points or more from this normative value would be important to detect. Based on historical data, he anticipates that the standard deviation of scores in his district is about 80 points.

The researcher sets alpha at 0.05 and would like to obtain power of 90%. He would also like to report the group mean with a precision (90% confidence interval) of plus or minus 15 points.

▶ Choose *One sample t-test that mean = specified value*.

For the following steps, see Figure 7.1:

▶ Enter the group names—*Local school district* and *Test against constant*.

▶ Enter the effect size—a population mean of 520 versus a constant of 500, with a standard deviation of 80.

▶ Click *Alpha* and select the following values: *Alpha = 0.05*, *Confidence Level = 95%*, and *Tails = 2*.

▶ Press Ctrl-F (or click the *Find N* icon if the default power has already been set to 90). The program shows that a sample of 171 students will yield power of 90%.

▶ The program shows that an observed mean of 520 would be reported, with a 95% confidence interval of 507.95 to 532.05.

▶ Click the *Report* icon to generate the following report:

Power for a test of the null hypothesis

One goal of the proposed study is to test the null hypothesis that the population mean is 500.0. The criterion for significance (alpha) has been set at 0.05. The test is two-tailed, which means that an effect in either direction will be interpreted.

With the proposed sample size of 171, the study will have power of 90.2% to yield a statistically significant result.

This computation assumes that the population from which the sample will be drawn has a mean of 520.0 with a standard deviation of 80.0. The observed value will be tested against a theoretical value (constant) of 500.0.

This effect was selected as the smallest effect that would be important to detect, in the sense that any smaller effect would not be of clinical or substantive significance. It is also assumed that this effect size is reasonable, in the sense that an effect of this magnitude could be anticipated in this field of research.

Precision for estimating the effect size

A second goal of this study is to estimate the mean in the population. Based on these same parameters and assumptions, the study will enable us to report the mean with a precision (95.0% confidence level) of plus or minus 12.04 points.

For example, an observed mean of 520.0 would be reported with a 95.0% confidence interval of 507.96 to 532.04.

The precision estimated here is the median precision. Precision will vary as a function of the observed standard deviation (as well as sample size), and in any single study will be narrower or wider than this estimate.

8 Paired T-Test

Selecting the Procedure

To display the available procedures, choose *New analysis* from the File menu.

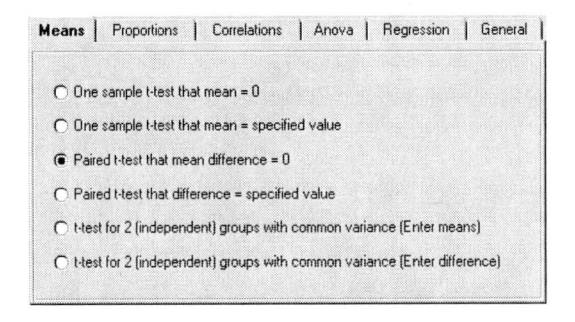

```
Means | Proportions | Correlations | Anova | Regression | General

  ○ One sample t-test that mean = 0
  ○ One sample t-test that mean = specified value
  ● Paired t-test that mean difference = 0
  ○ Paired t-test that difference = specified value
  ○ t-test for 2 (independent) groups with common variance (Enter means)
  ○ t-test for 2 (independent) groups with common variance (Enter difference)
```

Application

Figure 8.1 Paired t-test

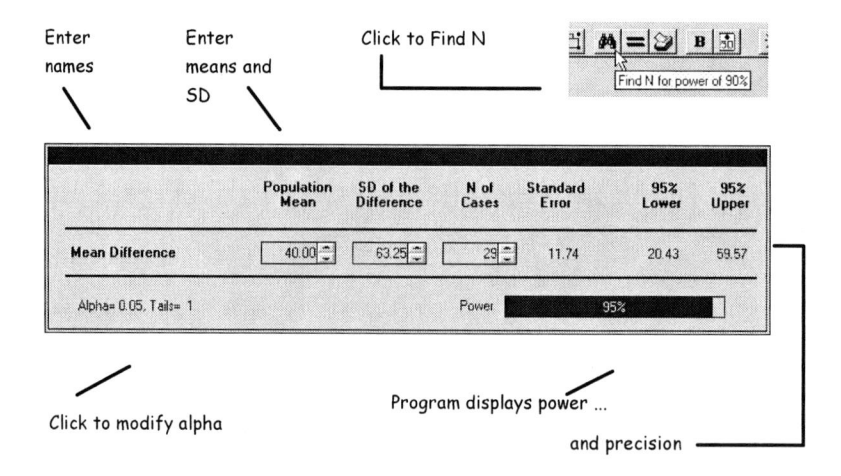

Enter names

Enter means and SD

Click to Find N

	Population Mean	SD of the Difference	N of Cases	Standard Error	95% Lower	95% Upper
Mean Difference	40.00	63.25	29	11.74	20.43	59.57

Alpha= 0.05, Tails= 1   Power   95%

Click to modify alpha

Program displays power ...

and precision

The paired t-test is used to compare two sets of scores when each score in one group is somehow linked with a score in the other group.

The test is commonly used to test a "difference" score, such as the change in a test score from one time point to the next. The paired t-test is also used in studies where persons in the two conditions are matched—for example, when a study enrolls pairs of siblings (one sibling assigned to each condition) or when patients are paired on the basis of disease severity before being assigned to either of two treatment plans.

It is always assumed that there will be a substantial positive correlation on the dependent variable between the two members of the pair. In other words, when we are testing a difference score, we assume that the pre-test and post-test are correlated. If this condition does not hold, it may be more appropriate (and powerful) to use a t-test for independent groups.

Effect Size

The effect size for t-tests is the standard difference, d, defined as the mean difference divided by the standard deviation of the difference.

In theory, the effect size index extends from 0 (indicating no effect) to infinity. In practice, of course, d is limited to a substantially smaller range, as reflected in the conventions suggested by Cohen for research in the social sciences—small (0.20), medium (0.50), and large (0.80).

Assume that patients are being entered into a study that will evaluate a treatment for cholesterol. Each patient's cholesterol level is assessed at baseline, the patient is put on a special regimen for six months, and the level is assessed again. The change from pre-treatment to post-treatment will be tested by a paired t-test. A clinically meaningful effect would be a drop of 40 points per patient over the treatment period.

In the paired t-test, the effect size is computed using the standard deviation of the difference, rather than the standard deviation at the pre-test (or post-test).

This is an important distinction. Assume, for example, that the treatment works as expected—every patient's cholesterol level drops by some 40 points. John enters treatment with a level of 400 and leaves with a level of around 360. Paul enters treatment with a level of 280 and leaves with a level of 240. If we look at the group as a whole, the treatment effect is obscured by the overlap between the two time points. Specifically, John's level *after* treatment is substantially worse (360) than Paul's level *prior to* treatment (280). By contrast, if we look at the treatment's impact on a patient-by-patient basis, we see reductions that consistently fall within a fairly narrow range of 40 points. Clearly, the second perspective yields a more compelling case that the treatment is effective.

The distinction between these two perspectives is the distinction between the test of two independent groups (one assessed prior to treatment and the other assessed after treatment) and a single group followed over time (which is the subject of this chapter). This distinction is made by using the standard deviation of the difference, rather than the standard deviation of either time point, to compute the effect.

If the researcher is able to estimate the standard deviation of the difference score, the program will accept this value directly. In many cases, the user will be able to estimate the standard deviation of the individual scores and the correlation between pre-treatment and post-treatment scores but may not be able to estimate the standard deviation of the difference. The program will accept the two standard deviations and the correlation, and compute the standard deviation of the difference. In the current example, the researcher, based on historical data, assumes that the pre/post correlation is about 0.80.

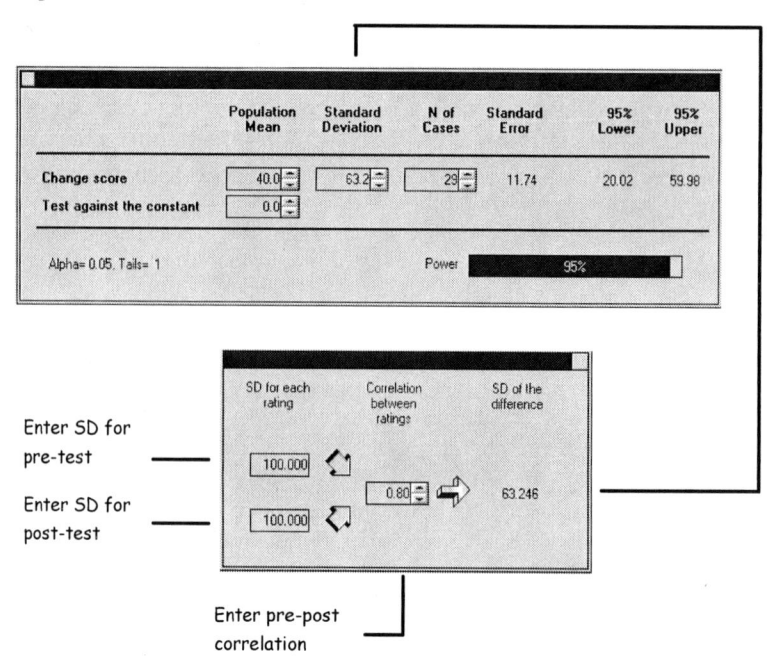

Figure 8.2 Standard deviation assistant

	Population Mean	Standard Deviation	N of Cases	Standard Error	95% Lower	95% Upper
Change score	40.0	63.2	29	11.74	20.02	59.98
Test against the constant	0.0					

Alpha= 0.05, Tails= 1   Power   95%

	SD for each rating	Correlation between ratings	SD of the difference
Enter SD for pre-test	100.000		
		0.80	63.246
Enter SD for post-test	100.000		

Enter pre-post correlation

The program includes two versions of the paired t-test. The first assumes that the difference will be tested against the nil value (0). In the running example, the user would enter 40 (which will be tested against 0). The second allows the user to test the mean against any specific value. For the running example, the user could enter 260 versus 300.

Tip

To enter the standard deviation of the difference directly, close the assistant box. When the assistant box is open (the default), the program requires that you enter the standard deviation for each time point and the correlation between the two sets of scores. These data are used to compute the standard deviation of the difference, which is transferred to the main panel.

Alpha, Confidence Level, and Tails

Click *Alpha*, *Tails*, or the confidence level to modify these values. The value set for tails applies to both the power analysis and the confidence interval.

Sample Size

The spin control adjacent to the N of Cases value can be used to modify the sample size quickly. The size of the increment can be specified by the user (click *N of Cases*).

Click the *Find N* icon to have the program find the number of cases required for the default level of power. The default value for power is 80%, but this can be modified temporarily (Ctrl-F) or permanently (choose *Preferences* from the Options menu).

Tolerance Interval

The confidence interval displayed by the program is the median interval and is shown balanced about the expected mean. For example, assume that the user enters an expected mean difference of 0, with SD Difference = 10 (the standard deviation for each time point is 10 and the pre/post correlation is 0.50) and Number pairs = 40. The program displays the 95% confidence interval as −3.17 to +3.17.

Choose *Tolerance intervals* from the View menu. The program shows that the 50% tolerance interval for the 95% confidence interval is 6.34 (corresponding to the distance from −3.15 to +3.15), which means that in 50% of the studies, the 95% confidence interval will be no wider than 6.34 points.

Additionally, it shows that the 80% tolerance interval is 6.96, which means that in 80% of all studies, the 95% confidence interval will be no wider than 6.96.

To modify the confidence level, click on the value shown, or choose *Alpha/Tails* from the Options menu. To modify the tolerance level, change the value shown in the tolerance interval box.

Because a one-tailed "interval" actually extends from minus infinity to some value above the mean, or from some value below the mean to plus infinity, it is not useful here. We could, however, consider the distance from the observed mean to the single boundary. If the user selects a one-tailed confidence interval, the median distance from the observed mean to the single boundary will be no greater than 2.64 in 50% of samples and no greater than 2.90 in 80% of samples.

As the sample size increases, not only are we able to estimate the mean more precisely (which yield a narrower confidence interval), but the dispersion of the sample standard deviation about the expected standard deviation decreases; this means that the width of the confidence interval shows less variation from sample to sample. For example, with Number pairs = 200, the 50% and 80% tolerance intervals are quite similar, at 2.78 and 2.90, respectively (assuming that SD = 10,0).

When the user specifies that the variance is known (choose *Computational formulas* from the Options menu), the confidence interval is computed using the population variance specified by the user. In this case, the confidence interval will not vary from sample to sample, and the 80% (or any other) tolerance interval is identical to the 50% tolerance interval.

Computational Options for Power

By default, the program assumes that the population variance is unknown (will be estimated based on the observed variance), which is normally the case. The program allows the user to specify that the population variance is known (technically, this would be a z-test rather than a t-test). In this case, the program will work with the z-distribution rather than the t-distribution for computing power and confidence intervals. In practice, this distinction will have a substantial impact only if the number of cases is quite small (under 20–30) and the effect size is large. With larger sample sizes, the impact on computations may not be evident.

To switch between the t-test and the z-test, choose *Computational formulas* from the Options menu.

This module applies exclusively to the paired t-test. The formulas used are identical to those used by the one-sample t-test module, but the reports and tools are optimized for each module. The present module should *not* be used for the two-sample t-test because the computational formulas are different.

Computational Options for Precision

The computational options for precision are identical to those for power. By default, the program works with the t-distribution (for estimated variance), but the user can choose to work with the z-distribution (for known variance). The option set for power is applied to precision as well.

Options for Screen Display

The program will display the standard difference, d. Choose *Customize screen* from the Options menu.

Example

This illustration is a continuation of the example used for effect size. The researcher will enroll patients whose cholesterol levels fall in the range of 200–450, have them modify their diets for a year, and then test their cholesterol levels again.

There is no logical reason for the diet to *increase* cholesterol levels. More to the point, the researcher's interest is limited to finding a *decrease* in levels. A finding that the diet increases levels would have the same substantive impact as a finding of no effect because either result would mean that the diet should not be adopted. On this basis, the researcher decides to use a one-tailed test.

Alpha will be set at 0.05 and the researcher wants the study to have power of 90%. Additionally, she wants to know the precision with which she will be able to report the effect size.

▶ Choose *Paired t-test that the mean difference = 0*.

For the following steps, see Figure 8.2:

▶ Enter the mean change—40 points.

▶ Enter the standard deviation for each time point (100) and the pre/post correlation (0.80). The program inserts the standard error of the difference (63.2).

▶ Click *Alpha* and select the following values: *Alpha = 0.05*, *Confidence Level = 95%*, and *Tails = 1*.

▶ Press Ctrl-F (or click the *Find N* icon if the default power has already been set to 95). The program shows that a sample of 29 patients will yield power of 95%.

▶ The program shows that an observed change of 40 points would be reported, with a 95% confidence interval of 20.26 to 59.74.

▶ Click the *Report* icon to generate the following report:

Power for a test of the null hypothesis

One goal of the proposed study is to test the null hypothesis that the population mean change is 0.0. The criterion for significance (alpha) has been set at 0.05. The test is one-tailed, which means that only an effect in the expected direction will be interpreted.

With the proposed sample size of 29, the study will have power of 95.3% to yield a statistically significant result.

This computation assumes that the population from which the sample will be drawn has a mean change of 40.0 with a standard deviation of 63.2. The observed value will be tested against a theoretical value (constant) of 0.0.

This effect was selected as the smallest effect that would be important to detect, in the sense that any smaller effect would not be of clinical or substantive significance. It is also assumed that this effect size is reasonable, in the sense that an effect of this magnitude could be anticipated in this field of research.

Precision for estimating the effect size

A second goal of this study is to estimate the mean change in the population. Based on these same parameters and assumptions, the study will enable us to report the mean change with a precision (95.0% confidence level) of plus or minus 19.74 points.

For example, an observed mean change of 40.0 would be reported with a 95.0% confidence interval of 20.43 to infinity or (alternatively, per the a priori hypothesis) minus infinity to 59.57. (Because the confidence interval has been defined as one-tailed, only one boundary is meaningful).

The precision estimated here is the median precision. Precision will vary as a function of the observed standard deviation (as well as sample size), and in any single study will be narrower or wider than this estimate.

9 T-Test for Independent Groups

Selecting the Procedure

To display the available procedures, choose *New analysis* from the File menu.

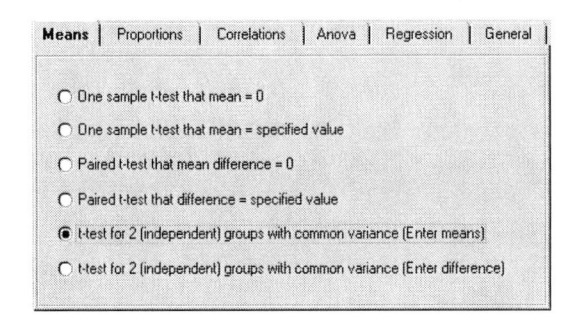

| **Means** | Proportions | Correlations | Anova | Regression | General |

○ One sample t-test that mean = 0
○ One sample t-test that mean = specified value
○ Paired t-test that mean difference = 0
○ Paired t-test that difference = specified value
● t-test for 2 (independent) groups with common variance (Enter means)
○ t-test for 2 (independent) groups with common variance (Enter difference)

Application

Figure 9.1 T-test for two independent groups

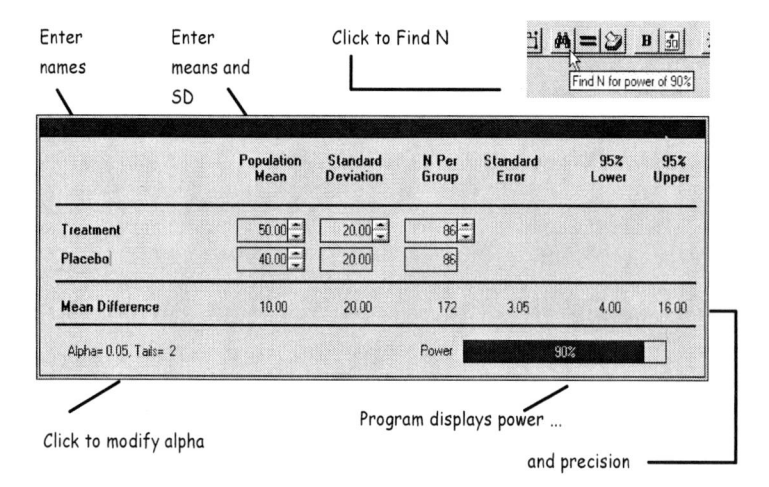

Enter names

Enter means and SD

Click to Find N

Find N for power of 90%

	Population Mean	Standard Deviation	N Per Group	Standard Error	95% Lower	95% Upper
Treatment	50.00	20.00	86			
Placebo	40.00	20.00	86			
Mean Difference	10.00	20.00	172	3.05	4.00	16.00

Alpha= 0.05, Tails= 2   Power   90%

Click to modify alpha

Program displays power ...

and precision

This procedure is used to test the mean difference in two independent groups. It is intended for a case in which the two groups share a common within-group standard deviation. The program allows the user to enter the mean difference directly or to enter the mean for each group (the program will display the mean difference).

Effect Size

The effect size for t-tests is the standard difference, d, defined as the mean difference between groups divided by the common within-group standard deviation.

Assume, for example, that two populations have mean scores on the SAT of 550 versus 500 and that the standard deviation within either population is 100 points. The effect size ((550 − 500)/ 100) is 0.50.

In theory, the effect size index extends from 0 (indicating no effect) to infinity. In practice, of course, d is limited to a substantially smaller range, as reflected in the conventions suggested by Cohen for research in the social sciences—small (0.20), medium (0.50) and large (0.80).

The user can choose to enter the mean and standard deviation (SD) for each group, in which case the program will compute the mean difference (Mean1–Mean). Alternatively, the user can choose to enter the mean difference directly.

The program computes power for a t-test based on common within-groups standard deviations. Therefore, the user enters the standard deviation for the first group only, and this value is applied to (and displayed for) both groups.

The program will also allow the user to enter a separate standard deviation for each group (to activate the SD box for the second group, click on it). In this case, the program will compute the common within-group standard deviation using a weighted average of the two estimates and will display this value. This option is available only if the user is entering data for each group (rather than entering the raw difference).

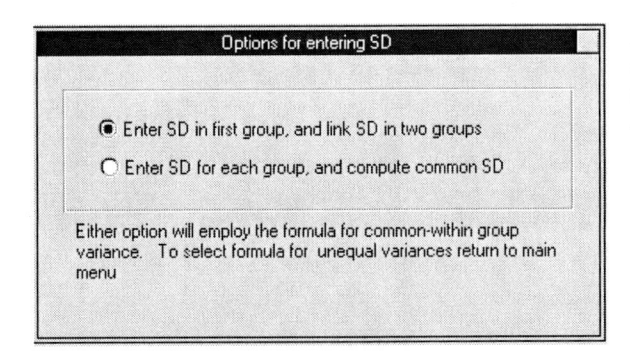

Activate from Options/ Data entry- Study design

Note. These alternatives affect the mode of data entry only, and *not* the computational formula. Both alternatives assume a common within-group standard deviation. The program does not compute power for the t-test based on different population variances.

Tip

To enter the standard difference (d) directly, set the program to enter the raw difference and provide a value of 1.0 for the standard deviation. In this case, the raw difference is equal to d.

Alpha, Confidence Level, and Tails

Click *Alpha*, *Tails*, or the confidence level to modify these values. The value set for tails applies to both the power analysis and the confidence interval.

Sample Size

The program assumes that cases will be entered into the two groups in equal numbers. To modify this assumption, click *N Per Group*. The program allows the user to link the cases using some other assignment ratio (for example, 1:2 or 3:5). As long as the groups are linked, the user will enter a sample size for one group only, which facilitates the process of finding an appropriate sample size. The program will also allow the user to enter the number for each group independently of the other.

The spin control adjacent to the N Per Group value can be used to modify the sample size quickly. The size of the increment can be specified by the user (click *N Per Group*).

Click the *Find N* icon to have the program find the number per group required for the default level of power. The program will honor any assignment ratio (for example, 2:1) that has been specified by the user. The default value for power is 90%, but this can be modified temporarily (Ctrl-F) or permanently (choose *Preferences* from the Options menu).

Tolerance Interval

The confidence interval displayed by the program is the expected, or average, interval and is shown balanced about the expected mean. For example, assume that the user enters an expected mean difference of 0, with SD = 10 for each group and N = 40 per group. The program displays the 95% confidence interval as −4.42 to 4.42.

Choose *Tolerance intervals* from the View menu. The program shows that the 50% tolerance interval for the 95% confidence interval is 8.84 (corresponding to the distance from −4.42 to +4.42), which means that in 50% of the studies, the 95% confidence interval will be no wider than 8.84 points.

Additionally, it shows that the 80% tolerance interval is 9.44, which means that in 80% of all studies, the 95% confidence interval will be no wider than 9.44.

To modify the confidence level, click on the value shown, or select *Alpha/Tails* from the Options menu. To modify the tolerance level, change the value shown in the tolerance interval box.

Because a one-tailed "interval" actually extends from minus infinity to some value above the mean, or from some value below the mean to plus infinity, it is not useful here. We could, however, consider the distance from the observed mean to the single boundary. If the user selects a one-tailed confidence interval, the 50% interval is shown as 7.39 and the 80% tolerance level is shown as 7.89. The user could report that the expected distance from the observed mean to the single boundary will be no greater than 3.70 (that is, half of the 7.39 interval) in 50% of cases and no greater than 3.95 (that is, half of the 7.89 interval) in 80% of cases.

As the sample size increases, not only are we able to estimate the mean more precisely (which yields a narrower confidence interval), but the dispersion of the sample standard deviation about the expected standard deviation decreases; this means that the width of the confidence interval shows less variation from sample to sample. For example, with N = 200 per group, the 50% and 80% tolerance intervals are quite similar to each other, at 3.29 and 3.39, respectively (assuming that SD = 10,0).

When the user specifies that the variance is known (choose *Computational formulas* from the Options menu), the confidence interval is computed using the population variance specified by the user. In this case, the confidence interval will not vary from sample to sample, and the 80% (or any other) tolerance interval is identical to the 50% tolerance interval.

Computational Options for Power

By default, the program assumes that the population variance is unknown and will be estimated based on the observed variance, which is normally the case. The program allows the user to specify that the population variance is known; technically, this would be a z-test rather than a t-test. In this case, the program will work with the z-distribution rather than the t-distribution for computing power and confidence intervals. In practice, this distinction will have an impact only if the number per group is quite small (under 20–30) and the effect size is large. With larger sample sizes, the impact on computations may not be evident.

To switch between the t-test and the z-test, choose *Computational formulas* from the Options menu.

This program assumes that the populations share a common within-group standard deviation. The program does not compute power for the t-test based on different population variances.

Computational Options for Precision

The computational options for precision are identical to those for power. By default, the program works with the t-distribution (for estimated variance), but the user can choose to work with the z-distribution (for known variance). The option set for power is applied to precision as well.

Options for Screen Display

The program will display the standard difference, d. Choose *Customize screen* from the Options menu.

Example

Patients suffering from allergies will be assigned at random to one of two treatment conditions (treatment versus placebo) and asked to rate their comfort level on a scale of 0 to 100. The expected standard deviation within groups is 20 points. The researcher believes that a 10-point difference between groups is the smallest difference that would be important to detect.

The researcher sets alpha at 0.05 and would like to obtain power of 90%. She would also like to report the size of the effect with a precision (90% confidence interval) of plus or minus 5 points.

▶ Choose *t-test for 2 (independent) groups with common variance (Enter means)*.

For the following steps, see Figure 9.1:

▶ Enter names for the two groups—*Treatment* and *Placebo*.

▶ Enter the mean for each group—50 and 40.

▶ Enter the standard deviation for the first group—20. This value is applied to the second group as well.

▶ Click *Alpha* and select the following values: *Alpha = 0.05*, *Confidence Level = 95%*, and *Tails = 2*.

▶ Press Ctrl-F (or click the *Find N* icon if the default power has already been set to 90). The program shows that a sample of 86 patients per group will yield power of 90%.

▶ The program shows that an observed difference of 10 points would be reported, with a 95% confidence interval of 3.99 to 16.01.

▶ Click the *Report* icon to generate the following report:

Power for a test of the null hypothesis

One goal of the proposed study is to test the null hypothesis that the two population means are equal. The criterion for significance (alpha) has been set at 0.05. The test is two-tailed, which means that an effect in either direction will be interpreted.

With the proposed sample size of 86 and 86 for the two groups, the study will have power of 90.3% to yield a statistically significant result.

This computation assumes that the mean difference is 10.00 (corresponding to a means of 50.00 versus 40.00) and that the common within-group standard deviation is 20.00.

This effect was selected as the smallest effect that would be important to detect, in the sense that any smaller effect would not be of clinical or substantive significance. It is also assumed that this effect size is reasonable, in the sense that an effect of this magnitude could be anticipated in this field of research.

Precision for estimating the effect size

A second goal of this study is to estimate the mean difference between the two populations. On average, a study of this design would enable us to report the mean difference with a precision (95.0% confidence level) of plus or minus 6.01 points.

For example, an observed difference of 10.00 would be reported with a 95.0% confidence interval of 3.99 to 16.01.

The precision estimated here is the median precision. Precision will vary as a function of the observed standard deviation (as well as sample size) and in any single study will be narrower or wider than this estimate.

10 Proportions in One Sample

Selecting the Procedure

To display the available procedures, choose *New analysis* from the File menu.

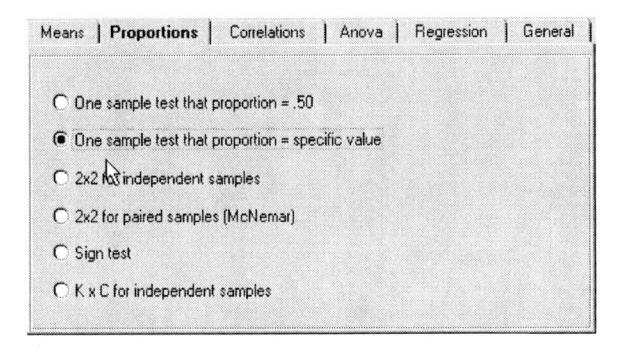

| Means | **Proportions** | Correlations | Anova | Regression | General |

- ○ One sample test that proportion = .50
- ◉ One sample test that proportion = specific value
- ○ 2x2 for independent samples
- ○ 2x2 for paired samples (McNemar)
- ○ Sign test
- ○ K x C for independent samples

Application

Figure 10.1 One-sample test that proportion equals .50

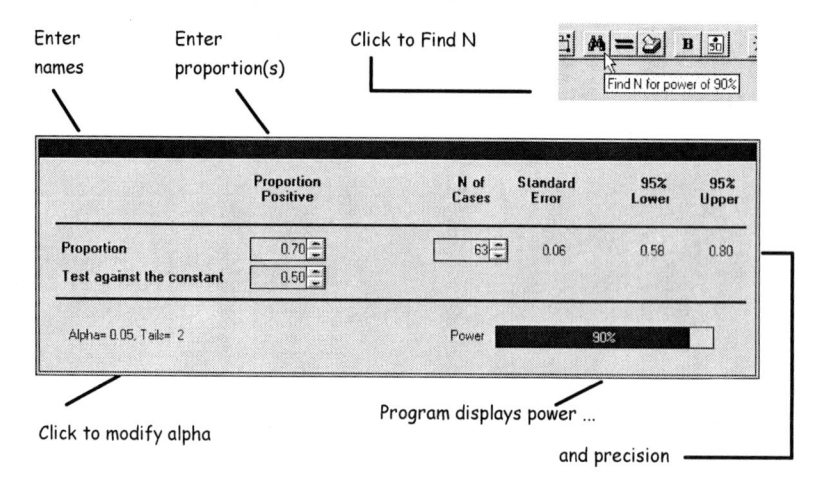

[Figure showing a dialog with annotations: "Enter names", "Enter proportion(s)", "Click to Find N", "Find N for power of 90%". Table columns: Proportion Positive, N of Cases, Standard Error, 95% Lower, 95% Upper. Row "Proportion" values: 0.70, 63, 0.06, 0.58, 0.80. "Test against the constant" 0.50. "Alpha= 0.05, Tails= 2", "Power 90%". Annotations: "Click to modify alpha", "Program displays power ... and precision"]

The program includes two versions of the one-sample test of proportion. The first version assumes that the difference will be tested against the null value of 50%. For example, patients are being treated with radiation, which carries the burden of serious side effects but the promise of a better long-term outcome. We want to survey patients after the fact and find out if the majority feel that the decision to have radiation was the correct one. (The null hypothesis is that 50% will respond positively and 50%, negatively.) A clinically important effect would be one that differed from 50% by 20 percentage points or more.

The second version allows the user to customize the null hypothesis. For example, we expect that 80% of patients are satisfied that they elected to have radiation, and we want to test the null hypothesis that 70% feel this way. In this case we would enter 0.80 for the proportion and 0.70 for the constant.

Effect Size

The effect size for the one-sample test of proportions is based on the difference between the two proportions. Unlike the t-test, where a difference of 10 versus 20 is equivalent to a difference of 40 versus 50 (that is, a 10-point difference in either case), when we work with proportions, the absolute values of the two proportions are relevant. Concretely, a difference of 10% versus 20% represents a more detectable effect than a difference of 40% versus 50%. (The variance of the effect size is larger, and the effect size is smaller, when we work with proportions near 0.50.)

For this reason, we refer to the effect by citing the rate difference followed by the actual values—for example, a 10-point difference (40% versus 50%).

The effect size reported in this way can range from 0 (indicating no effect) to 0.999. For research in the social sciences, Cohen has suggested the following conventional values—small (0.55 versus 0.50), medium (0.65 versus 0.50), and large (0.75 versus 0.50).

Alpha, Confidence Level, and Tails

Click *Alpha*, *Tails*, or the confidence level to modify these values. The value set for tails applies to both the power analysis and the confidence interval.

Sample Size

The spin control adjacent to the N of Cases value can be used to modify the sample size quickly. The size of the increment can be specified by the user (click *N of Cases*).

Click the *Find N* icon to have the program find the number of cases required for the default level of power. The default value for power is 90%, but this may be modified temporarily (Ctrl-F) or permanently (select *Preferences* from the Options menu).

Note. The Find N procedure is available only when the program is using the normal (arcsin) approximation to compute power.

Computational Options for Power

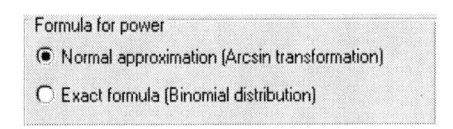

Formula for power
- Normal approximation (Arcsin transformation)
- Exact formula (Binomial distribution)

The program offers two methods for computing power: the arcsin transformation, which is an approximation, and the exact formula, which employs the binomial distribution.

The binomial formula is recommended for small sample sizes (under 30), and the arcsin formula is recommended for larger samples (as the number of cases grows, the arcsin method approaches the accuracy of the exact method and has the advantage of being faster).

When the user chooses to work with the binomial test for power, the program displays the actual alpha for the test. This is always less than or equal to the alpha specified by the user, and it varies as a function of sample size.

78   Chapter 10

To switch between computational methods, select *Computational formulas* from the Options menu.

Note. The Find N function is available only for the arcsin procedure. When the binomial formula is in effect, a small increase in sample size may result in a drop in power. This apparent anomaly is due to the fact that power is based on actual alpha rather than nominal alpha, and the increase in N may result in a more conservative actual alpha.

Computational Options for Precision

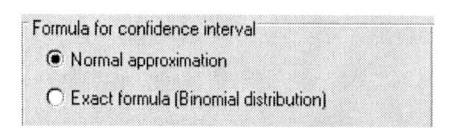

```
Formula for confidence interval
  ⦿ Normal approximation
  ○ Exact formula (Binomial distribution)
```

The program offers two options for computing confidence intervals: a method based on the normal approximation and an exact method based on the binomial formula. To switch between computational methods, select *Computational formulas* from the Options menu.

Example

A researcher wants to test the null hypothesis that 50% of patients will report that their quality of life has improved as a result of their treatment. The smallest difference that would be important to detect is a difference of 70% versus this null value.

▶ Choose *One sample test that proportion = specific value*.

For the following steps, see Figure 10.1:

▶ Choose either version of the One Proportion test.

▶ For effect size, enter 0.70 (versus 0.50).

▶ Click *Alpha* and select the following values: *Alpha = 0.05*, *Confidence Level = 95%*, and *Tails = 2*.

▶ Press Ctrl-F (or click the *Find N* icon if the default power has already been set to 90). The program shows that a sample of 63 patients will yield power of 90%.

▶ The program shows that an observed proportion of 70% would be reported, with a 95% confidence interval of 58% to 80%.

Power for a test of the null hypothesis

One goal of the proposed study is to test the null hypothesis that the proportion positive in the population is 0.50. The criterion for significance (alpha) has been set at 0.05. The test is two-tailed, which means that an effect in either direction will be interpreted.

With the proposed sample size of 63, the study will have power of 90.4% to yield a statistically significant result.

This computation assumes that the proportion positive in the population is 0.70. The observed value will be tested against a theoretical value (constant) of 0.50

This effect was selected as the smallest effect that would be important to detect, in the sense that any smaller effect would not be of clinical or substantive significance. It is also assumed that this effect size is reasonable, in the sense that an effect of this magnitude could be anticipated in this field of research.

Precision for estimating the effect size

A second goal of this study is to estimate the proportion positive in the population. Based on these same parameters and assumptions, the study will enable us to report the this value with a precision (95.0% confidence level) of approximately plus or minus 0.11.

For example, an observed proportion of 0.70 would be reported with a 95.0% confidence interval of 0.58 to 0.80.

The precision estimated here is the expected (average) value over many studies. Precision will vary as a function of the observed proportion (as well as sample size) and in any single study will be narrower or wider than this estimate.

▶ Click the *Report* icon to generate this report:

11 Proportions in Two Independent Groups

Selecting the Procedure

To display the available procedures, choose *New analysis* from the File menu.

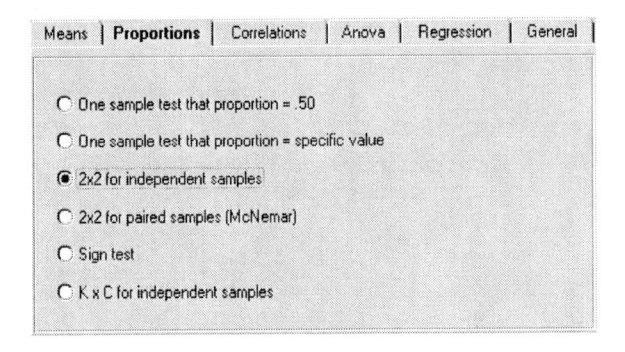

| Means | **Proportions** | Correlations | Anova | Regression | General |

- ○ One sample test that proportion = .50
- ○ One sample test that proportion = specific value
- ● 2x2 for independent samples
- ○ 2x2 for paired samples (McNemar)
- ○ Sign test
- ○ K x C for independent samples

Application

Figure 11.1 2x2 for independent samples

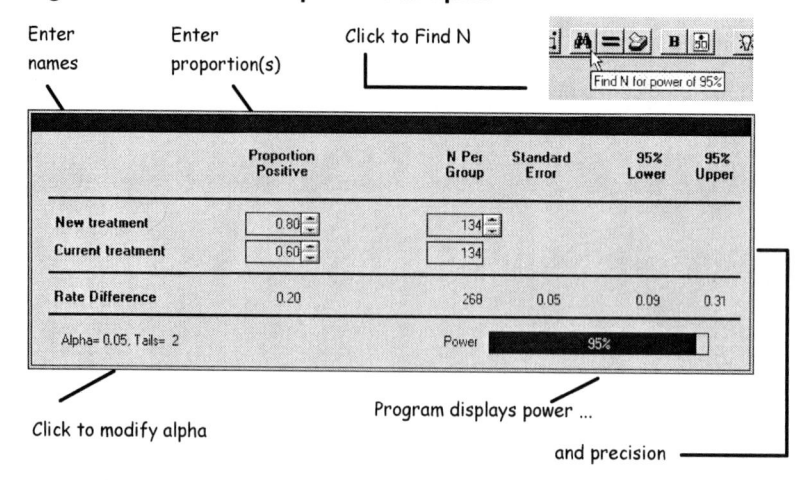

The two-group test of proportions is used to test the hypothesis that the proportion of cases meeting some criterion is identical in the two groups.

For example, we are planning to assign patients to one of two treatment options and then test the null hypothesis that the treatments are equally effective (that is, that the proportion of patients cured will be identical in the two populations).

The user's attention is called to the fact that the program assumes that the row marginals are fixed. This assumption is not likely to be reasonable for certain types of studies, such as case control studies. This can have important implications for the computation of power as well as precision.

Effect Size

The effect size for the two-sample test of proportions is based on the difference between the two proportions. Unlike the t-test, where a difference of 10 versus 20 is equivalent to a difference of 40 versus 50 (that is, a 10-point difference in either case), when we work with proportions, the absolute values of the two proportions are relevant. Concretely, a difference of 10% versus 20% represents a more detectable effect than a difference of 40% versus 50%.

For this reason, we refer to the effect by citing the rate difference followed by the actual values—for example, a 10-point difference (40% versus 50%).

The effect size reported in this way can range from 0 (indicating no effect) to 0.999. For research in the social sciences, Cohen has suggested the following conventional values—small (40% versus 50%), medium (40% versus 65%), and large (40% versus 78%).

Alpha, Confidence Level, and Tails

Click *Alpha*, *Tails*, or the confidence level to modify these values. The value set for tails applies to both the power analysis and the confidence interval.

Sample Size

The program assumes that cases will be entered into the two groups in equal numbers. To modify, click *N of Cases*. The program allows the user to link the cases using some other assignment ratio (for example, 1:2 or 3:5). As long as the groups are linked, the user will enter a sample size for one group only. The program also allows the user to enter the number for each group independently of the other.

The spin control adjacent to the N of Cases value can be used to modify the sample size quickly. The size of the increment can be specified by the user (click *N of Cases*).

Click the *Find N* icon to have the program find the number of cases required for the default level of power. The program will honor any assignment ratio (for example, 2:1) that has been specified by the user. The default value for power is 90%, but this can be modified temporarily (Ctrl-F) or permanently (choose *Preferences* from the Options menu).

Computational Options for Power

The program allows the user to select from several computational formulas for power. To switch between options, choose *Computational formulas* from the Options menu.

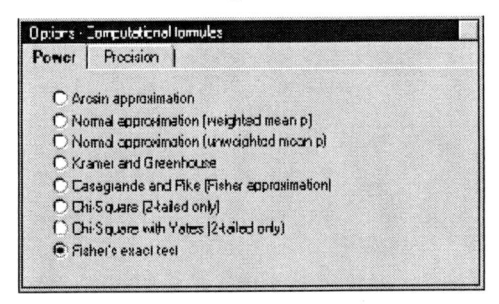

The most accurate estimate of power is given by Fisher's exact test (this is true even if the study will be analyzed by the chi-square test rather than the Fisher test), but this computation (unlike the others) involves an iterative procedure.

The user is advised to use the Casagrande and Pike option, which yields an excellent approximation to Fisher's exact test. This method requires no iterations and allows access to all of the program tools (Find N, tables, etc.). As a final step, for a small number of cases, you may want to select *Fisher's exact test* to get an exact computation, which will generally fall within 0.01 or 0.02 of the Casagrande estimate.

The other computational options are summarized here. These options became popular because they do not require a computer, but they have little to recommend them here. With large sample sizes, all formulas yield similar results.

Normal approximation. Two options for the normal approximation are offered: unweighted and weighted. Under the former, the null value is set at the midpoint between the two populations; under the latter, it is weighted by the number of cases in either group. When the two groups are of equal size, the two formulas are identical. Except under special circumstances, the unweighted option should be selected. The program also includes the arcsin, a variant on the normal approximation discussed by Cohen (1988).

Chi-square. The program allows the user to compute power using the noncentral chi-square distribution, with or without the Yates correction. These options are available for a two-tailed test only.

Approximations to Fisher's exact test. A number of approximations have been developed for computing power for Fisher's exact test. The Kramer-Greenhouse method is seen as overly conservative and generally yields the lowest estimates of power. The Casagrande and Pike method yields results that are very close to the exact computation of power by the Fisher method.

The Fisher exact method. The program allows the user to compute power for Fisher's exact test. This test gives the most accurate estimate of power, even when the post-study analysis will be performed using the chi-square test rather than the Fisher test.

Computational Options for Precision

The program offers two options for computing confidence intervals: the log method or the Cornfield method.

Options for Screen Display

By default, the program displays the rate difference corresponding to the two proportions. The program will also compute the corresponding relative risk and odds ratio and display these together with the corresponding expected confidence intervals.

This feature allows the researcher to ensure that the effect size specified for power corresponds to a relative risk (or odds ratio) that is appropriate. Additionally, if the researcher intends to report the relative risk or odds ratio after the study is completed, this provides an estimate of the precision with which these parameters will be reported. To toggle the display of these indices, choose *Customize screen* from the Options menu.

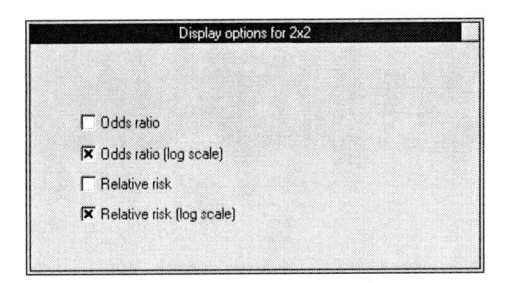

Example

We are planning to assign patients to one of two treatment options (new versus current) and to test the null hypothesis that the treatments are equally effective (that is, that the proportion of patients cured will be identical in the two populations).

The cure rate for the current treatment is approximately 60%. We anticipate that the new treatment will prove more effective in curing the illness. However, the new treatment is also expected to have more serious side effects. For this reason, the treatment would be recommended only if it increased the cure rate by 20 percentage points. Accordingly, the effect size selected for the power analysis will be 60% versus 80%.

While it is unlikely that the aggressive treatment will result in a cure rate lower than the standard, this possibility cannot be ruled out entirely, and it would have practical implications for future research. For this reason, the analysis will be two-tailed.

▶ Choose *2x2 for independent samples*.

For the following steps, see Figure 11.1:

▶ Enter the group names—*New treatment* and *Current treatment*. Click *Proportion Positive* and change it to *Proportion Cured*.

▶ Enter the effect size—proportions of 0.80 and 0.60.

▶ Click *Alpha* and select the following values: *Alpha = 0.05*, *Confidence Level = 95%*, and *Tails = 2*.

▶ Press Ctrl-F (or click the *Find N* icon if the default power has already been set to 95). The program shows that a sample of 134 patients per group will yield power of 95%.

▶ The program shows that an observed rate difference of 20 points would be reported with a 95% confidence interval of 9 points to 31 points. To narrow the confidence interval, the user could increase the sample size (or modify the confidence level).

▶ Click the *Report* icon to generate the following report:

Power for a test of the null hypothesis

One goal of the proposed study is to test the null hypothesis that the proportion positive is identical in the two populations. The criterion for significance (alpha) has been set at 0.05. The test is two-tailed, which means that an effect in either direction will be interpreted.

With the proposed sample size of 134 and 134 for the two groups, the study will have power of 95.1% to yield a statistically significant result.

This computation assumes that the difference in proportions is 0.20 (specifically, 0.80 versus 0.60).

This effect was selected as the smallest effect that would be important to detect, in the sense that any smaller effect would not be of clinical or substantive significance. It is also assumed that this effect size is reasonable, in the sense that an effect of this magnitude could be anticipated in this field of research.

Precision for estimating the effect size

A second goal of this study is to estimate the difference between the two populations. Based on these same parameters and assumptions, the study will enable us to report the difference in proportions with a precision (95.0% confidence level) of approximately plus or minus 0.11.

Specifically, an observed difference of 0.20 would be reported with a 95.0% confidence interval of 0.09 to 0.31.

The precision estimated here is the approximate expected value over many studies. Precision will vary as a function of the observed proportions (as well as sample size) and, in any single study, will be narrower or wider than this estimate.

12 Paired Proportions

Selecting the Procedure

To display the available procedures, choose *New Analysis* from the File menu.

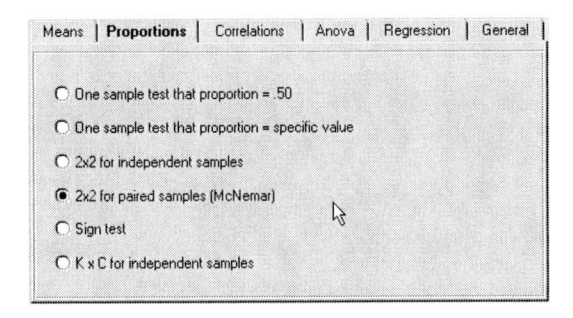

| Means | **Proportions** | Correlations | Anova | Regression | General |

- ○ One sample test that proportion = .50
- ○ One sample test that proportion = specific value
- ○ 2x2 for independent samples
- ● 2x2 for paired samples (McNemar)
- ○ Sign test
- ○ K x C for independent samples

Application

Figure 12.1 2X2 for paired samples (McNemar)

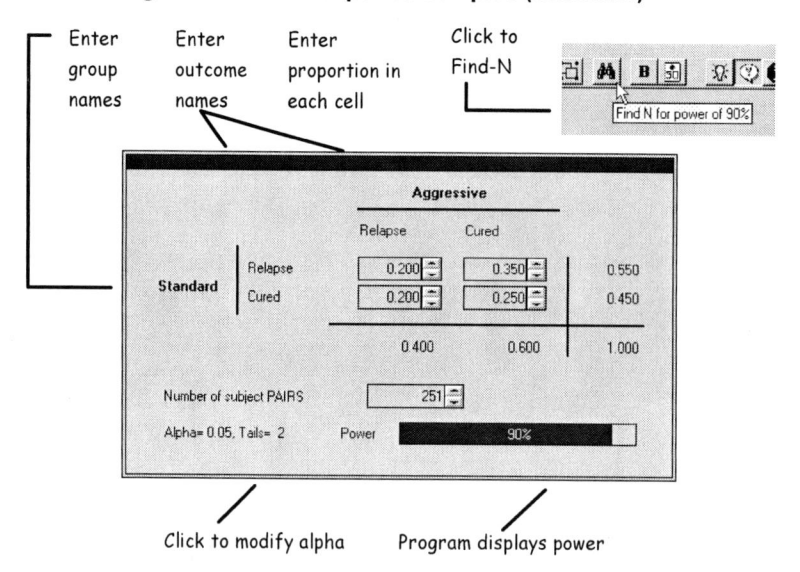

Enter group names — Enter outcome names — Enter proportion in each cell — Click to Find-N

		Aggressive		
		Relapse	Cured	
Standard	Relapse	0.200	0.350	0.550
	Cured	0.200	0.250	0.450
		0.400	0.600	1.000

Number of subject PAIRS: 251
Alpha= 0.05, Tails= 2   Power 90%

Click to modify alpha — Program displays power

The McNemar test of paired proportions is used to compare the proportion of cases in two groups when the cases in the two groups are matched in a way that is relevant to the outcome—for example, when patients are matched on the basis of disease severity and then assigned to one of two treatments, or when siblings are assigned to one of two conditions.

Effect Size

The effect size for the two-sample test of paired proportions is based on the difference between the two proportions. In this sense, the McNemar test is identical to the test of proportions in two independent groups. In the case of the McNemar test, however, we work with pairs of cases rather than individual cases.

Assume, for example, that patients are matched for disease severity and then assigned to either the standard treatment or to an aggressive treatment. The possible outcomes are *sick* or *cured*. When both members of a pair are classified as sick or both are classified as cured, the pair provides no information about the relative utility of the two treatments. Therefore, the test of the hypothesis that the two treatments are equally effective is based entirely on cases where one member of the pair was cured and the other was not.

It follows that for the purpose of computing power, the effective number of pairs is the number falling in the upper right and lower left cells of a 2x2 table (that is, those cells where one member of the pair is cured and the other is not). For the purpose of computing power, the effect size is the difference between the proportion in these two cells.

As is true for any test of proportions, a difference of 10% versus 20% is *not* equivalent to a difference of, say, 40% versus 50%, despite the fact that the difference in either case is 10 percentage points. (Proportions near 50% have larger variance and therefore smaller effect sizes). For this reason, we refer to the effect by citing the rate difference followed by the actual values—for example, a 10-point difference (40% versus 50%).

Alpha and Tails

Click *Alpha* or *Tails* to modify these values.

Sample Size

The program requires the user to enter a value for *Total Number of Pairs* and then specify the proportion of pairs expected to fall into each cell. As above, the number of pairs for computation of power will be determined by the program based on the product of these two values.

For any given number of cases, power will be higher if the proportion of cases falling into the upper right and lower left cells is increased, which increases the effective number of cases. Of course, power will also increase as the disparity between the upper right and lower left cells increases, since this is the basis of the effect size.

The spin control adjacent to the *N of Cases* value can be used to modify the sample size quickly. The size of the increment can be specified by the user (click *N of Cases*).

Click the *Find N* icon to have the program find the number of cases required for the default level of power. The default value for power is 90%, but this can be modified temporarily (Ctrl-F) or permanently (select *Preferences* from the Options menu).

Computational Options for Power

The program allows the user to select the normal (arcsin) approximation or the exact (binomial) test. When the sample size is small, the binomial test is preferable. Otherwise, the normal approximation should be selected (with a large sample size, the binomial option can require several seconds for iterations). To select either option, choose *Computational formulas* from the Options menu.

Example

We are planning to assign patients to match patients on the basis of disease severity and assign the two members of each pair to separate treatment options (standard versus aggressive). We will test the null hypothesis that the treatments are equally effective (in that the proportion of patients cured will be identical in the two populations).

We anticipate that in 20% of pairs, both patients will relapse; in 25%, both will be cured; in 35%, the patient in the aggressive group will have the better outcome; in 20%, the patient in the standard group will have the better outcome.

While it is unlikely that the aggressive treatment will result in a cure rate lower than the standard, this possibility cannot be ruled out entirely, and it would have practical implications for future research. For this reason, the analysis will be two-tailed.

▶ Choose *2x2 for paired samples (McNemar)*.

For the following steps, see Figure 12.1:

▶ Enter group names—*Aggressive* and *Standard*.

▶ Enter names for the outcomes—*Relapse* and *Cured*.

▶ Enter the proportion falling into each cell (0.20, 0.35, 0.20, and 0.25).

▶ Click *Alpha* and select the following values: *Alpha = 0.05, Confidence Level = 95%, Tails = 2*.

▶ Press Ctrl-F (or click the *Find N* icon if the default power has already been set to 90). The program shows that a sample of 251 pairs (502 patients) will yield power of 90%.

▶ Click the *Report* icon to generate the following report:

Power for a test of the null hypothesis

One goal of the proposed study is to test the null hypothesis that there is no relationship between the variable used to classify cases (Aggressive versus Standard) and outcome (Relapse versus Cured). This will be tested by the McNemar test for paired proportions. The criterion for significance (alpha) has been set at 0.05 (two-tailed).

The power analysis is based on the following population effect size: in 20.0% of pairs, both cases will be classified as Relapse and in another 25.0%, both cases will be classified as Cured. These cases contribute no information to the hypothesis test.

A discrepancy between the two members of a pair is expected in the balance of the population: 35.0% of all pairs will show an outcome of Cured for the Aggressive case only, while 20.0% of all pairs will show an outcome of Cured for the Standard case only. This effect was selected as the smallest effect that would be important to detect, in the sense that any smaller effect would not be of clinical or substantive significance. It is also assumed that this effect size is reasonable, in the sense that an effect of this magnitude could be anticipated in this field of research.

With the proposed sample size of 251, the study will have power of 90.1% to yield a statistically significant result.

13 Sign Test

Selecting the Procedure

To display the available procedures, choose *New analysis* from the File menu.

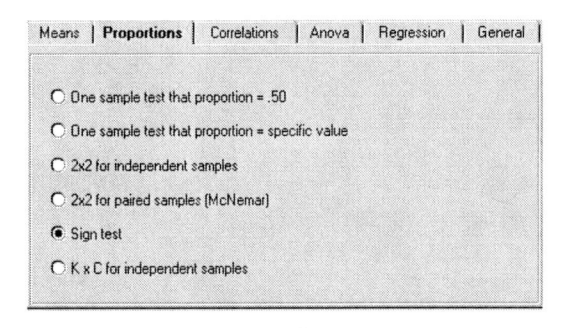

Means	**Proportions**	Correlations	Anova	Regression	General

○ One sample test that proportion = .50
○ One sample test that proportion = specific value
○ 2x2 for independent samples
○ 2x2 for paired samples (McNemar)
⦿ Sign test
○ K x C for independent samples

Application

Figure 13.1 Sign test

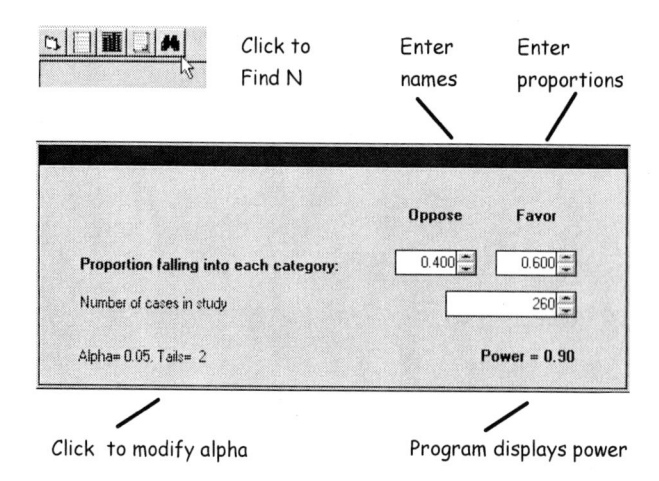

Click to Find N — Enter names — Enter proportions

	Oppose	Favor
Proportion falling into each category:	0.400	0.600
Number of cases in study		260
Alpha= 0.05, Tails= 2		Power = 0.90

Click to modify alpha — Program displays power

The sign test is used to compare the proportion of cases in one of two mutually exclusive groups. For example, we may classify people as planning to vote for one of two candidates.

In this test, the proportion of cases falling into either outcome is completely determined by the proportion in the complementary group. Therefore, the user is required to enter the proportion falling into either one of the groups only. A proportion of, say, 40% in one outcome group implies a proportion of 60% in the other.

Effect Size

The effect size for the two-sample test of paired proportions is based on the difference between either proportion and the null hypothesis of 50% (of necessity, the two proportions are equidistant from this null value).

For research in the social sciences, Cohen has suggested the following conventional values—small (0.45 versus 0.55), medium (0.35 versus 0.65), and large (0.25 versus 0.75). Wherever possible, however, the selection of an effect size should be based on the research issues, that is, substantively.

Alpha and Tails

Click *Alpha* or *Tails* to modify these values.

Sample Size

The program requires the user to enter a total number of cases and then specify the proportion falling into either of the two outcomes.

The spin control adjacent to the N of Cases value can be used to modify the sample size quickly. The size of the increment can be specified by the user (click *N of Cases*).

Click the *Find N* icon to have the program find the number of cases required for the default level of power. The default value for power is 90%, but this can be modified temporarily (Ctrl-F) or permanently (choose *Preferences* from the Options menu).

Computational Options for Power

The program allows the user to select the arcsin approximation or the exact (binomial) test. When the sample size is small, the binomial test is preferable. Otherwise, the normal approximation should be selected (with a large sample size, the binomial option can require several seconds for iterations). To select either option, choose *Computational formulas* from the Options menu.

Example

A politician wants to conduct a poll to find out how her constituents feel about an issue on the upcoming ballot. If 60% of the population has a preference, it would be important to reject the null hypothesis of no effect.

▶ Choose *Sign test*.

For the following steps, see Figure 13.1:

▶ Enter the group names—*Oppose* and *Favor*.

▶ Enter the proportion falling into the *Favor* cell (0.60).

▶ Click *Alpha* and select the following values: *Alpha = 0.05* and *Tails = 2*.

▶ Press Ctrl-F (or click the *Find N* icon if the default power has already been set to 90). The program shows that a sample of 260 persons will yield power of 90%.

▶ Click the *Report* icon to generate the following report:

Power for a test of the null hypothesis

One goal of the proposed study is to test the null hypothesis that Oppose and Favor contain equal proportions of cases (that is, that half of the population falls into either classification). The criterion for significance (alpha) has been set at 0.05 (two-tailed).

The power analysis is based on a population effect size such that 40% of cases fall into Oppose and 60% fall into Favor, for a discrepancy of 20 percentage points. This effect was selected as the smallest effect that would be important to detect, in the sense that any smaller effect would not be of clinical or substantive significance. It is also assumed that this effect size is reasonable, in the sense that an effect of this magnitude could be anticipated in this field of research.

With the proposed sample size of 260, the study will have power of 90.1% to yield a statistically significant result.

14 K x C Crosstabulation

Selecting the Procedure

To display the available procedures, choose *New analysis* from the File menu.

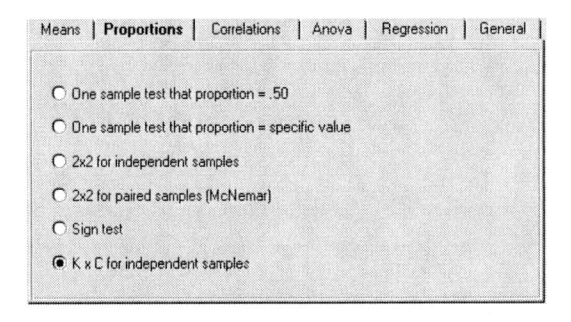

| Means | **Proportions** | Correlations | Anova | Regression | General |

- ○ One sample test that proportion = .50
- ○ One sample test that proportion = specific value
- ○ 2x2 for independent samples
- ○ 2x2 for paired samples (McNemar)
- ○ Sign test
- ● K x C for independent samples

# Application

Figure 14.1   K x C crosstabulation

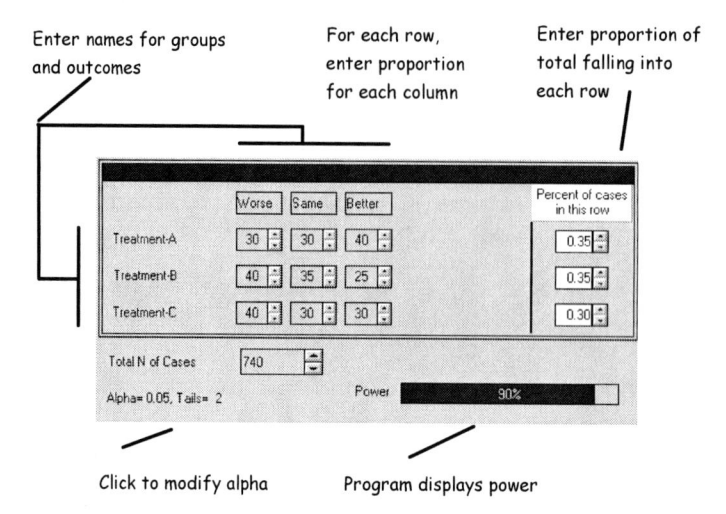

The K × C (K rows by C columns) crosstabulation is used to test the hypothesis that classification on one dimension (for example, assignment to one of three treatments) is related to classification on another dimension (for example, outcome, where patients are assigned to one of three possible outcomes).

Effect Size

This discussion will refer to a running example in which patients are assigned to one of three treatment groups and then classified as having one of three outcomes. However, the test is not limited to situations in which one variable logically precedes the other. Additionally, the test does not require that the matrix of rows by columns be square.

Under the null hypothesis of no effect, outcome will have no relation to treatment. If 10% of patients are classified as improved, then this number should be 10% for each of the three treatment groups. If 40% of patients are classified as the same, then this number should be 40% for each of the treatment groups.

The effect size, w, is based on the disparity between the matrix of proportions that we would expect under the null hypothesis and the matrix that is assumed under the alternative hypothesis.

To enter the effect size for the K × C crosstabulation:

▶ Specify the number of rows and columns in the table (choose *Data entry/Study design* from the Options menu).

▶ Enter the percentage of cases falling into each column for each row in the table. The proportions for each row must sum to 100.

▶ Enter the proportion of all cases falling into each row of the table. The proportions must sum to 1.00.

This approach is somewhat different from the approach sometimes taken in texts, in which the researcher is required to enter the proportion of the full population falling into each cell (so that *all* cells, rather than the cells in a single row, sum to 1.00). The approach taken by this program ensures that the cells for two rows will appear to be identical under the null, even if more cases are assigned to one row than another. This makes it easier for the researcher to identify the places in which the rows will differ.

The effect size computed in this way (w) is analogous to the chi-square value used to test for significance, but is a pure measure of effect size, whereas chi-square is affected also by the sample size. The relation between the two is given by:

chi-square = w^2 * Ntotal

The effect size, w, can be used to derive two additional indices of effect—the contingency coefficient (C) and Cramér's phi. The program can be set to display these values as well as w.

Alpha

Click *Alpha* to modify its value.

Sample Size

The program assumes initially that cases will be divided evenly across the rows (50% to each of two groups, 33.3% to each of three groups, and so on), but this can be modified by the user. For example, the user can specify that 50% of all cases will be assigned to *Treatment A* and 25% each to *Treatments B* and *C*.

The spin control adjacent to the N of Cases value can be used to modify the sample size quickly. The size of the increment can be specified by the user (click *N of Cases*).

Click the *Find N* icon to have the program find the number of cases required for the default level of power. The default value for power is 90%, but this can be modified temporarily (Ctrl-F) or permanently (choose *Preferences* from the Options menu).

Computational Options for Power

Power is computed with reference to the noncentral chi-square distribution. No options are available.

Example

We are planning to assign patients to one of three treatment options. Subsequent to treatment, each patient will be classified as *Worse*, *Same*, or *Better*. The null hypothesis is that the treatments are equally effective.

▶ Choose *K x C for independent samples*.

For the following steps, see Figure 14.1:

▶ From the Options menu choose *Data entry/Study design* and set 3 rows and 3 columns.

▶ Enter the group names—*Treatments A, B,* and *C*.

▶ Enter the names for the outcomes—*Worse, Same,* and *Better*.

▶ For the first row, enter the percentage falling into each cell (30, 30, and 40). Repeat for the second (40, 35, and 25) and third (40, 30, and 30) rows.

▶ Enter the proportion of the full sample assigned to each of the three treatments—0.35, 0.35, and 0.30.

▶ Click *Alpha* and select the following value: *Alpha = 0.05*.

▶ Press Ctrl-F (or click the *Find N* icon if the default power has already been set to 90). The program shows that a sample of 740 patients will yield power of 90%.

▶ Click the *Report* icon to generate the following report:

Power for a test of the null hypothesis

One goal of the proposed study is to test the null hypothesis that the proportion of cases falling into each column is identical for all rows in the study. The criterion for significance (alpha) has been set at 0.05 (two-tailed).

Treatment A cases will be distributed as 30% in Worse, 30% in Same, and 40% in Better. These cases represent 35% of the sample.

Treatment B cases will be distributed as 40% in Worse, 35% in Same, and 25% in Better. These cases represent 35% of the sample.

Treatment C cases will be distributed as 40% in Worse, 30% in Same, and 30% in Better. These cases represent 30% of the sample.

With the proposed sample size of 740, the study will have power of 90.0% to yield a statistically significant result.

15 Correlation—One Group

Selecting the Procedure

To display the available procedures, choose *New analysis* from the File menu.

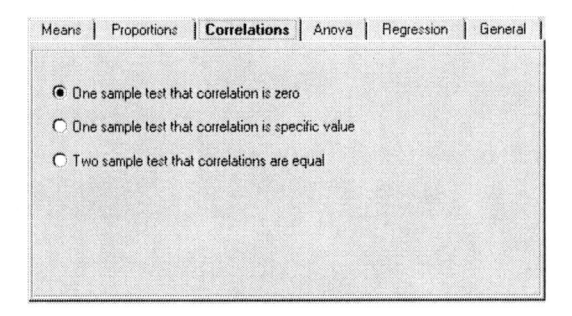

- ◉ One sample test that correlation is zero
- ○ One sample test that correlation is specific value
- ○ Two sample test that correlations are equal

Application

Figure 15.1 One-sample test that correlation is 0

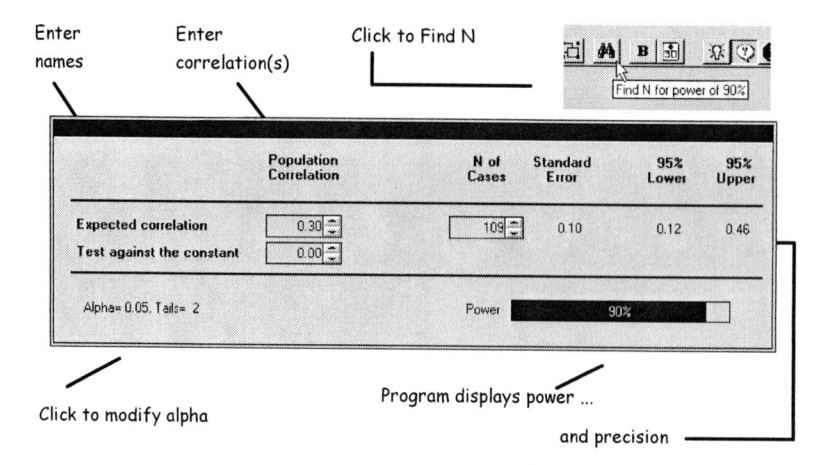

```
Enter          Enter              Click to Find N
names          correlation(s)
```

	Population Correlation	N of Cases	Standard Error	95% Lower	95% Upper
Expected correlation	0.30	109	0.10	0.12	0.46
Test against the constant	0.00				

Alpha= 0.05, Tails= 2   Power 90%

Click to modify alpha

Program displays power ... and precision

The one-sample correlation procedure is used to test the hypothesis that a correlation is 0 (or any other specific value).

Effect Size

When a correlation is being tested against the null value of 0, the effect size is simply the correlation coefficient (r). The effect size, r, can range from 0 to an absolute value of 1.00 (the program will accept any value up to 0.999). The program allows r to be either negative or positive, but the effect size is based on the absolute value.

The program will also compute power for a test of the null other than 0. The effect size is based on the difference between the two correlations (the null value and the correlation under the alternate), and the sign of each correlation is important. For correlations, an effect size of 0.30 versus 0.10 is not the same as an effect size of 0.50 versus 0.30, despite the fact that the difference is 20 points in either case.

Alpha, Confidence Level, and Tails

Click *Alpha, Tails*, or the confidence level to modify these values. The value set for tails applies to both the power analysis and the confidence interval.

Sample Size

The spin control adjacent to the *N of Cases* value can be used to modify the sample size quickly. The size of the increment can be specified by the user (click *N of Cases*).

Click the *Find N* icon to have the program find the number of cases required for the default level of power. The default value for power is 90%, but this can be modified temporarily (Ctrl-F) or permanently (choose *Preferences* from the Options menu).

Computational Options for Power

Power for a correlation versus the null of 0 is computed using exact methods (similar to those used for multiple regression).

Power for testing a single correlation versus a constant other than 0 is carried out by means of the Fisher-Z transformation.

Example 1

A college is trying out a test that it may use to place students into sections of a mathematics class in the future. At present, all students take the test during the first week of class and then attend the same class for the duration of the semester. The hypothesis is that scores on the placement test will be correlated with scores on the final examination.

The smallest correlation that would be meaningful is a correlation of 0.30, so this is the value used for effect size in computing power. The college decides to run the study with alpha set at 0.05 and power at 90%. While the test is not expected to have a negative correlation with grades, a finding that it does would have important implications, thus the test would be two-tailed.

▶ Choose *One sample test that correlation is zero*.

For the following steps, see Figure 15.1:

▶ Enter the correlation of 0.30.

▶ Click *Alpha* and select the following values: *Alpha = 0.05* and *Tails =2*.

▶ Press Ctrl-F (or click the *Find N* icon). The program shows that a sample of 109 students will yield power of 90%.

▶ The program shows also that an observed correlation of 0.30 would be reported with a 95% confidence interval of 0.12 to 0.46.

▶ Click the *Report* icon to generate the following report:

Power for a test of the null hypothesis

One goal of the proposed study is to test the null hypothesis that the correlation in the population is 0.00. The criterion for significance (alpha) has been set at 0.05. The test is two-tailed, which means that an effect in either direction will be interpreted.

With the proposed sample size of 109, the study will have power of 90.2% to yield a statistically significant result.

This computation assumes that the correlation in the population is 0.30. The observed value will be tested against a theoretical value (constant) of 0.00

This effect was selected as the smallest effect that would be important to detect, in the sense that any smaller effect would not be of clinical or substantive significance. It is also assumed that this effect size is reasonable, in the sense that an effect of this magnitude could be anticipated in this field of research.

Precision for estimating the effect size

A second goal of this study is to estimate the correlation in the population. Based on these same parameters and assumptions, the study will enable us to report the this value with a precision (95.0% confidence level) of approximately plus/minus 0.17 points.

For example, an observed correlation of 0.30 would be reported with a 95.0% confidence interval of 0.12 to 0.46.

The precision estimated here is the approximate value expected over many studies. Precision will vary as a function of the observed correlation (as well as sample size), and in any single study will be narrower or wider than this estimate.

Example 2

The situation is identical to Example 1. In this case, however, the administrator decides that it will be of little value to know merely that the correlation exceeds 0. The smallest correlation that would be important as a useful predictor is 0.30, and he wants to test the null hypothesis that the true correlation is 0.30.

If we can assume that the true correlation is 0.50, how many students would be needed to ensure power (90%) to reject this null hypothesis? Other parameters remain as before (alpha = 0.05, test is two-tailed).

▶ Choose *One sample test that correlation is specific value*.

For the following steps, see Figure 15.1:

▶ Enter the correlation of 0.50 for the population, to be tested against a constant of 0.30.

▶ Click *Alpha* and select the following values: *Alpha* = *0.05* and *Tails* =2.

▶ Press Ctrl-F (or click the *Find N* icon if the default power has already been set to 90). The program shows that a sample of 186 students will yield power of 90%.

▶ The program shows also that an observed correlation of 0.50 would be reported with a 95% confidence interval of 0.38 to 0.60 (note the lack of symmetry).

▶ Click the *Report* icon to generate the following report:

Power for a test of the null hypothesis

One goal of the proposed study is to test the null hypothesis that the correlation in the population is 0.30. The criterion for significance (alpha) has been set at 0.05. The test is two-tailed, which means that an effect in either direction will be interpreted.

With the proposed sample size of 186, the study will have power of 90.0% to yield a statistically significant result.

This computation assumes that the correlation in the population is 0.50. The observed value will be tested against a theoretical value (constant) of 0.30.

This effect was selected as the smallest effect that would be important to detect, in the sense that any smaller effect would not be of clinical or substantive significance. It is also assumed that this effect size is reasonable, in the sense that an effect of this magnitude could be anticipated in this field of research.

Precision for estimating the effect size

A second goal of this study is to estimate the correlation in the population. Based on these same parameters and assumptions, the study will enable us to report this value with a precision (95.0% confidence level) of approximately plus or minus 0.11 points.

For example, an observed correlation of 0.50 would be reported with a 95.0% confidence interval of 0.38 to 0.60.

The precision estimated here is the approximate value expected over many studies. Precision will vary as a function of the observed correlation (as well as sample size) and in any single study will be narrower or wider than this estimate.

16 Correlation—Two groups

Selecting the Procedure

To display the available procedures, choose *New analysis* from the File menu.

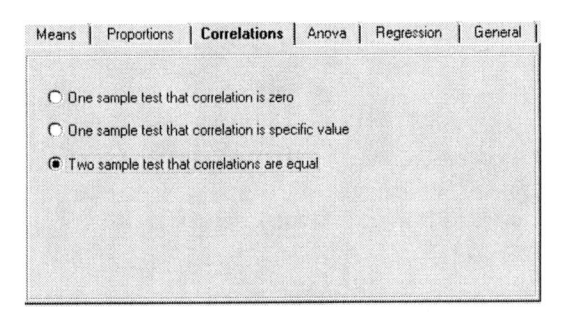

| Means | Proportions | **Correlations** | Anova | Regression | General |

- ○ One sample test that correlation is zero
- ○ One sample test that correlation is specific value
- ● Two sample test that correlations are equal

Application

Figure 16.1 Two-sample test that correlations are equal

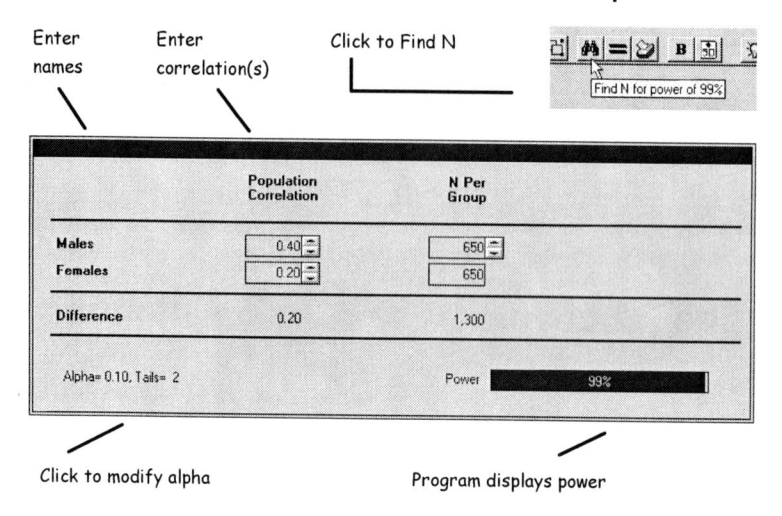

```
Enter          Enter              Click to Find N
names          correlation(s)
```

	Population Correlation	N Per Group
Males	0.40	650
Females	0.20	650
Difference	0.20	1,300

Alpha= 0.10, Tails= 2   Power   99%

Click to modify alpha   Program displays power

The two-sample correlation procedure is used to test the hypothesis that the correlation between X and Y is identical in populations A and B—for example, to test that the predictive utility of a screening test (indexed by the correlation between the test and a criterion) is identical for males and females.

Effect Size

The effect size used here is based on the difference between the two correlations coefficients. However, a difference of 0.20 versus 0.40 is *not* as detectable as a difference of 0.40 versus 0.60, despite the fact that the difference is 20 points in both cases. Therefore, the difference will be expressed by presenting the absolute difference followed by the two correlations.

The effect size, r, can range from the case where the two correlations are identical (indicating no effect) to the case where one correlation is −0.999 and the other is +0.999. When we are comparing two correlations, the sign of each value is, of course, important.

Alpha, Confidence Level, and Tails

Click *Alpha*, *Tails*, or the confidence level to modify these values.

Sample Size

The spin control adjacent to the *N of Cases* value can be used to modify the sample size quickly. The size of the increment can be specified by the user (click *N of Cases*).

Click the *Find N* icon to have the program find the number of cases required for the default level of power. The default value for power is 90%, but this can be modified temporarily (Ctrl-F) or permanently (select *Preferences* from the Options menu).

Computational Options for Power

Power is computed using the Fisher-Z transformation.

Example

A college is trying out a test that it may use to place students into sections of a mathematics class in the future. At present, all students take the test during the first week of class and then attend the same class for the duration of the semester. The hypothesis is that scores on the placement test will correlate with scores on the final examination.

It is anticipated that the correlation between the placement test and score on the final grade will be about 0.30. There is concern, however, that the ability of this test to function as a predictor of outcome will be substantially stronger for males, who as a group have had more formal training in mathematics, than for the females, who may have had less formal training.

If indeed the test functions as a strong predictor for males but not for females, it would be unfair to use this as a placement test for the whole group. The administration decides that the test will be considered to have this problem if the correlation between the placement test and the final grade is 20 points higher for one gender than for the other (0.40 for males versus 0.20 for females). The study's power should be 99%—if there really is a gender gap of 20 points, the study should have power of 99% to yield a significant effect.

▶   Choose *Two sample test that correlations are equal.*

For the following steps, see Figure 16.1.

▶ Enter the correlations of 0.40 and 0.20.

▶ Click *Alpha* and select the following values: *Alpha = 0.10, Tails =2*.

▶ Press Ctrl-F (or click the *Find N* icon if the default power has already been set to 99). The program shows that a sample of 650 students per group will yield power of 99%.

▶ Click the *Report* icon to generate the following report:

Power for a test of the null hypothesis

One goal of the proposed study is to test the null hypothesis that the correlation is identical in the two populations. The criterion for significance (alpha) has been set at 0.10. The test is two-tailed, which means that an effect in either direction will be interpreted.

With the proposed sample size of 650 and 650 for the two groups, the study will have power of 99.0% to yield a statistically significant result.

This computation assumes that the difference in correlations is 0.20 (specifically, 0.40 versus 0.20)

This effect was selected as the smallest effect that would be important to detect, in the sense that any smaller effect would not be of clinical or substantive significance. It is also assumed that this effect size is reasonable, in the sense that an effect of this magnitude could be anticipated in this field of research.

17 Analysis of Variance/Covariance (Oneway)

Selecting the Procedure

To display the available procedures, choose *New analysis* from the File menu.

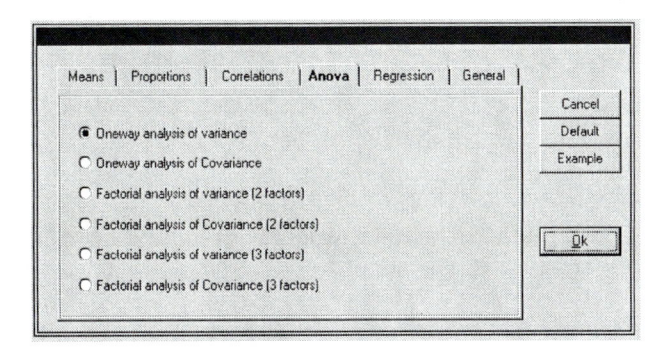

Application

Oneway analysis of variance (ANOVA) is used to compare the means in more than two groups. Assume, for example, that students are assigned at random to one of four classroom methods, taught for a period of time, and then assessed on some continuous measure. Oneway ANOVA might be used to test the null hypothesis of no difference in effect among the four groups.

Figure 17.1 Oneway analysis of variance

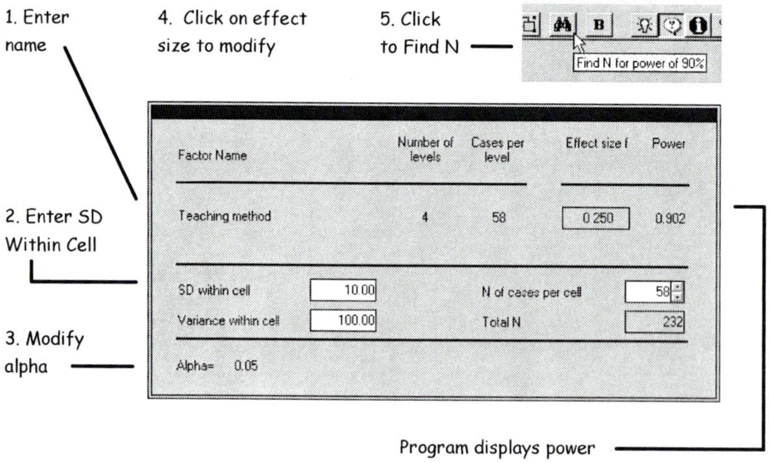

1. Enter name
2. Enter SD Within Cell
3. Modify alpha
4. Click on effect size to modify
5. Click to Find N

Program displays power

Effect Size

The effect size (f) used in analysis of variance is an extension of the effect size (d) used for a t-test. Recall that d is the mean difference between groups divided by the dispersion within groups. Similarly, f is based on the dispersion (standard deviation) between groups divided by the dispersion (standard deviation) within groups. (The effect size, f, is a pure measure of effect size and should not be confused with the F statistic, which takes into account sample size as well as effect size.)

The program allows the user to enter f directly or by using Cohen's conventions for research in the social sciences (small = 0.10, medium = 0.25, and large = 0.40).

Alternatively, the user is allowed to provide data that the program uses to compute f. In this case, the program requires the user to provide the within-cells standard deviation on the main screen. Then, enter data for the between-groups standard deviation in one of three formats:

- Enter the range of means for the factor.
- Enter the mean for each group.
- Enter the between-groups standard deviation or variance.

The program will use the data provided in any of these formats to compute the between-groups standard deviation and then proceed to compute f.

Entering the Effect Size (f) for Oneway ANOVA

▶ On the main screen, enter the *SD within cell* value (optional if the user will enter f directly).

Figure 17.2 Oneway ANOVA main screen

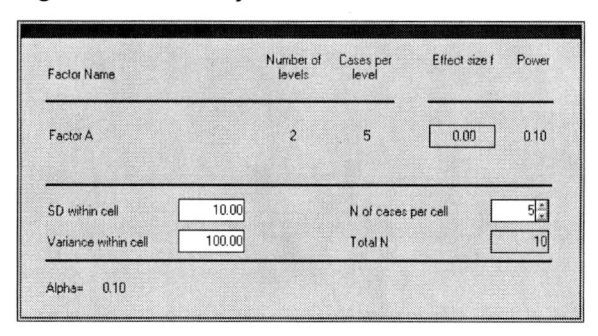

Factor Name	Number of levels	Cases per level	Effect size f	Power
Factor A	2	5	0.00	0.10

SD within cell 10.00   N of cases per cell 5
Variance within cell 100.00   Total N 10
Alpha= 0.10

▶ Click on the effect size shown (initially, this is 0.00). The program will immediately transfer control to one of four panels (see Figure 17.3).

116   Chapter 17

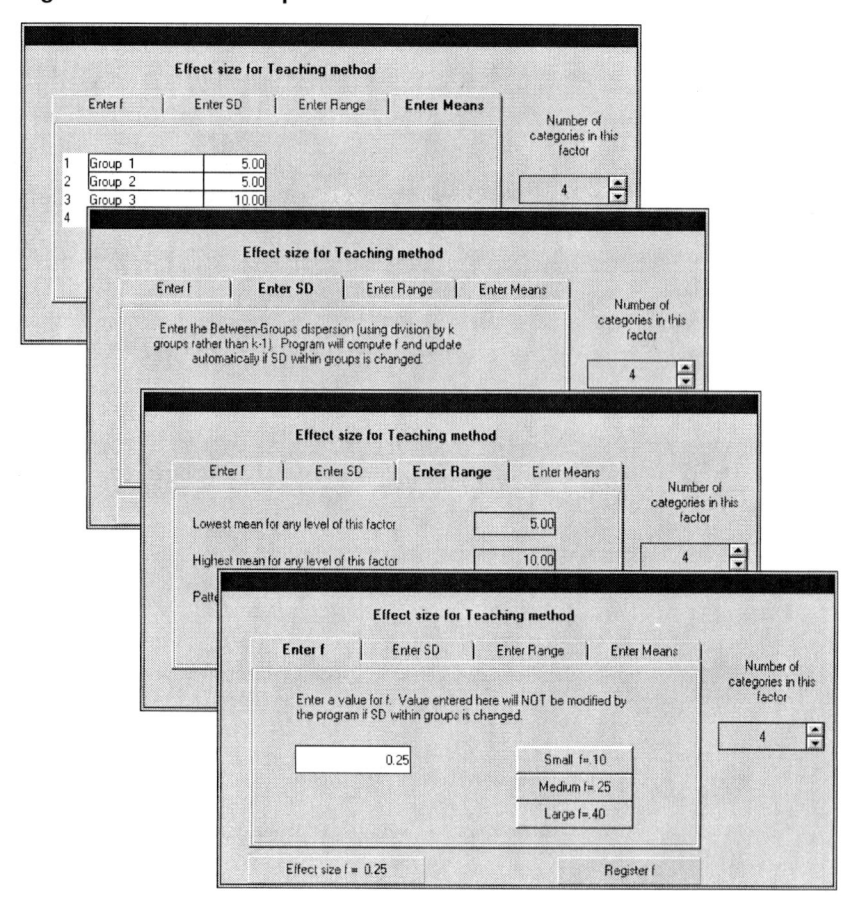

Figure 17.3   Effect size panels

▶ On the panel, enter the number of levels and the effect size.

Analysis of Variance/Covariance (Oneway) 117

Figure 17.4   Panel to enter f directly or using conventional values

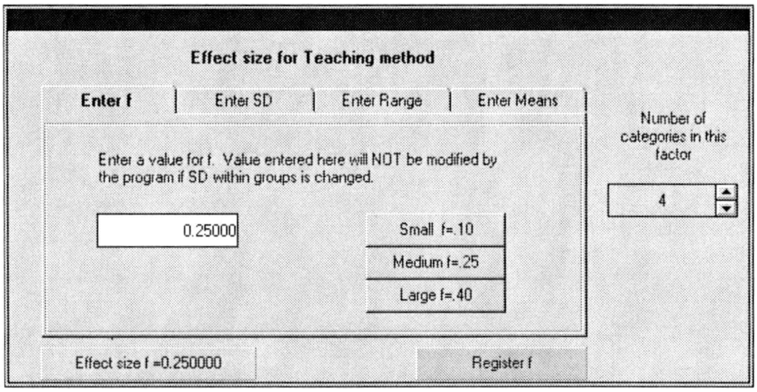

[Effect size for Teaching method panel: Enter f tab selected, with value 0.25000. Small f=.10, Medium f=.25, Large f=.40. Number of categories in this factor: 4. Effect size f = 0.250000. Register f button.]

This panel is intended for users who are familiar with the effect size, f, and feel comfortable specifying the f value directly. It is also appropriate for users who have little basis for estimating the effect size and therefore prefer to work with Cohen's conventions for small, medium, or large effects.

▶ Enter the number of groups.

▶ Click on one of the conventional values for effect size or enter a value directly. For example, a value of 0.25 would fall in the medium range according to Cohen's conventions.

▶ Click *Compute f* to compute the corresponding f.

▶ Click *Register f* to transfer the value to the main screen.

Figure 17.5 Panel to enter between-groups standard deviation

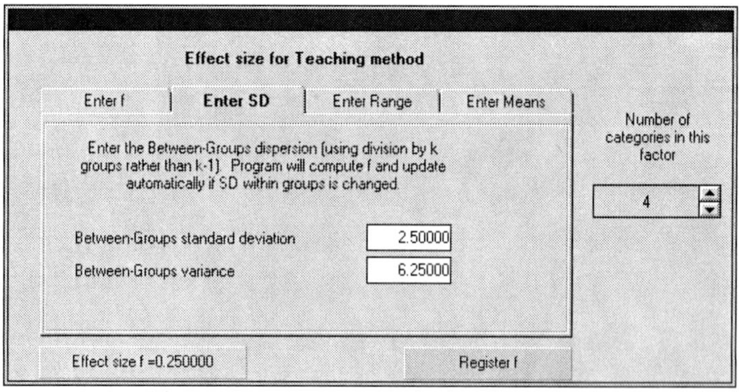

This panel is appropriate for researchers who are able to provide an estimate of the between-groups dispersion.

▶ Enter the number of groups.

▶ Enter the between-groups dispersion (either the standard deviation or the variance).

▶ Click *Compute f* to compute the corresponding f.

▶ Click *Register f* to transfer the value to the main screen.

In this example, the user entered the between-groups standard deviation (2.5) and the between-groups variance (6.25). Using this information and the within-cells standard deviation (10, entered on the main screen), the program computes the effect size, f (in this case, 2,5/10 = 0,25).

Analysis of Variance/Covariance (Oneway) 119

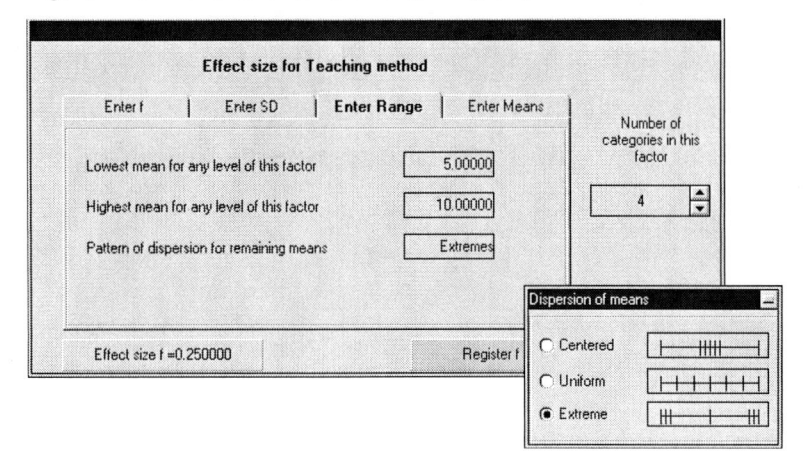

Figure 17.6 Panel to enter the range of group means

This panel is appropriate for researchers who are able to estimate the range of means but not the mean for each group.

▶ Enter the number of groups.

▶ Enter the single lowest and highest means.

▶ Once the two extreme groups have been specified, the remaining groups can fall into one of three patterns: all remaining groups fall at the center of the range (*Centered*), which will yield the smallest effect; the remaining groups are distributed evenly across the range (*Uniform*); or the remaining groups fall at either extreme (*Extreme*). Click on the value shown to activate the pop-up box and make a selection.

▶ Click *Compute f* to compute the corresponding f.

▶ Click *Register f* to transfer the value to the main screen.

Note that when the study has only two groups, the three patterns are identical and, thus, will have no effect on f.

On the main screen, the user specified that the within-cells standard deviation is 10. In Figure 17.6, the user specified that the four groups will have means that range from 5 to 10 and that the remaining two groups will have means at either extreme of this range (that is, at 5 and 10). The program has computed the corresponding f as 0.25.

Figure 17.7 Panel to enter the mean for each group

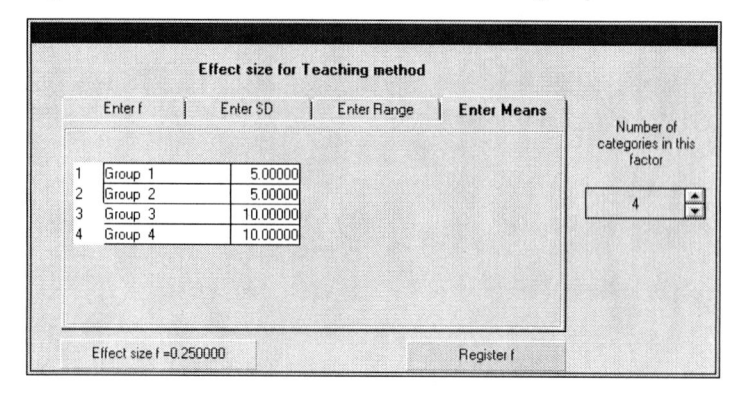

This panel is appropriate for researchers who are able to provide an estimate for every one of the group means.

▶ Enter the number of groups. The program will display an entry box for each group.

▶ Enter a name (optional) and mean (required) for each group.

▶ Click *Compute f* to compute the corresponding f.

▶ Click *Register f* to transfer the value to the main screen.

In this example, the user specified mean values of 5, 5, 10, and 10. These values provide the between-groups standard deviation (2.5), and the user has already provided the within-cells standard deviation (10). The program computes the effect size, f, as 0.25.

Correspondence between the Four Approaches

The four methods provided for computing f are mathematically equivalent to each other. In the example, the user is working with means of 5, 5, 10, and 10. The dispersion can be described by specifying these four distinct values. It can also be described by entering the range (5 to 10) and the pattern (extreme) of the two remaining means. Finally, it can be described by entering the standard deviation for the four means (2.5).

Each of these approaches yields the identical value for the between-groups standard deviation (2.5). Given the within-cells standard deviation (10) from the main screen, all yield the identical value for f (0.25).

Of course, the different methods yield the same value only when provided with equivalent information. Often, researchers will want to estimate f using more than one method as a check that the estimate is accurate. The program will retain the data entered in any panel, but only the registered data will be transferred to the main screen.

Effect Size Updated Automatically

When you initially enter the between-groups standard deviation (either by entering the between-groups standard deviation, the range of means, or the four individual means), the program will compute f using this value and the within-cells standard deviation (which is entered on the main screen). If you later modify the within-cells standard deviation, the effect size will be adjusted automatically to reflect this change (just as the effect size, d, is modified in a t-test when you update the within-groups standard deviation).

However, if you have decided to enter the effect size by specifying f directly, the effect size is not updated. The program assumes that a user who has specified a medium effect (0.25) may not know whether the between-groups/within-groups standard deviation is 0.25/1 or 2.5/10, but merely that a medium effect would be important to detect. In this case, the value entered for the within-cells standard deviation has no impact on f.

Alpha

Click *Alpha* to modify this value. ANOVA is sensitive to an effect in any direction and as such is nondirectional.

Sample Size

To modify the sample size, use the spin control. By default, this will increase the number of cases by 5, but this can be modified to any number (choose *N-Cases* from the Options menu).

Click the *Find N* icon to have the program find the number of cases required for the default level of power. The default value for power is 90%, but this can be modified temporarily (Ctrl-F) or permanently (choose *Preferences* from the Options menu).

Example 1

Students are assigned at random to one of four teaching groups, taught for a period of time, and then assessed on a continuous measure. A substantive effect would be a medium effect of f = 0.25.

▶ Choose *Oneway analysis of variance*.

▶ Enter the name *Teaching group*.

▶ Click on the value shown for effect size (0.00).

▶ Activate the *Enter f* tab. Set the number of groups (4) and effect size (medium).

▶ (Alternatively, enter the within-cells standard deviation on main screen and the between-groups standard deviation using the alternative panel.)

▶ Click *Compute f* and then click *Register f* to return to the main screen.

▶ Click *Alpha* and select the following value: *Alpha = 0.05*.

▶ Click the *Find N* icon (or press Ctrl-F). The program shows that 58 cases per cell will yield power of 90%.

▶ Click the *Report* icon to generate the following report:

Power for a test of the null hypothesis

This power analysis is for a oneway fixed effects analysis of variance with four levels. The study will include 58 cases per cell, for a total of 232 cases.

The criterion for significance (alpha) has been set at 0.05. The analysis of variance is nondirectional (that is, two-tailed), which means that an effect in either direction will be interpreted.

Main effects

Teaching group will include four levels, with 58 cases per level. The effect size (f) is 0.250, which yields power of 0.90.

Oneway Analysis of Covariance

Analysis of covariance is identical to analysis of variance except for the presence of the covariate. The covariates are able to explain some of the variance in the outcome measure, and by taking this into account, we reduce the error term, which yields a more powerful test.

We will illustrate the use of this procedure by extending the previous example. As in Example 1, students are assigned at random to one of three training groups, taught for a period of time, and then assessed on a continuous measure. In the ANOVA example, we simply assessed the students at the end point. For the ANCOVA example, we will assume that the students are assessed at the baseline as well and that the baseline ratings serve as a covariate.

The specification of effect size by the user proceeds exactly as that for ANOVA. Specifically, the within-groups standard deviation and the between-groups standard deviation are entered as though no covariate is present. If the user chooses to enter f directly, enter f as though no covariate is present. The covariate serves to reduce the error term (which boosts the effect size, f), but this is taken into account by the program, which displays both the unadjusted f and the adjusted f.

Figure 17.8 shows the oneway analysis of covariance. We assume that the baseline scores account for 40% of the scores at the end point.

Figure 17.8 Oneway analysis of covariance

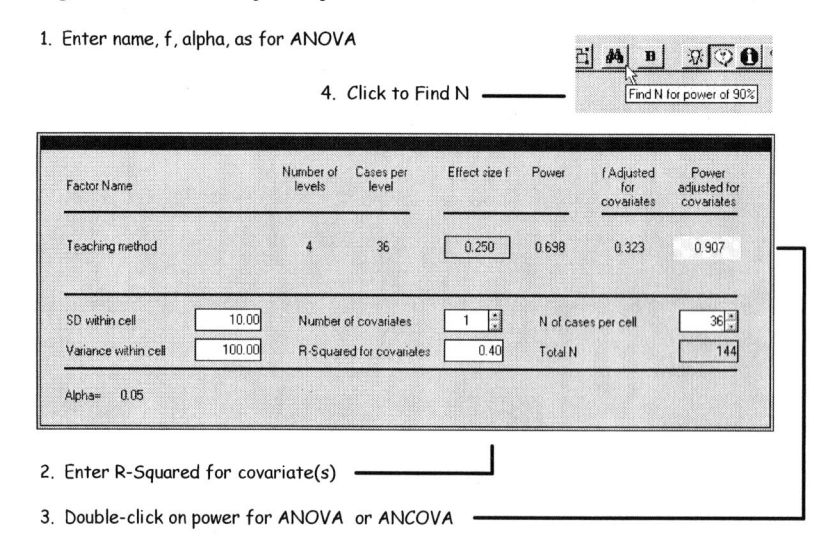

1. Enter name, f, alpha, as for ANOVA

4. Click to Find N

Factor Name	Number of levels	Cases per level	Effect size f	Power	f Adjusted for covariates	Power adjusted for covariates
Teaching method	4	36	0.250	0.698	0.323	0.907

SD within cell: 10.00 Number of covariates: 1 N of cases per cell: 36
Variance within cell: 100.00 R-Squared for covariates: 0.40 Total N: 144
Alpha= 0.05

2. Enter R-Squared for covariate(s)
3. Double-click on power for ANOVA or ANCOVA

Example 2

Students are assigned at random to one of four teaching groups, taught for a period of time, and then assessed on a continuous measure. A substantive effect would be a medium effect of f = 0.25. The pre-score will serve as a covariate and is expected to account for 40% of the variance in post-scores.

▶ Choose *Oneway analysis of covariance*.

▶ Enter the name *Teaching method*.

▶ Click on the value shown for effect size (0.00).

▶ Activate the *Enter f* tab. Set the number of groups (4) and effect size (medium).

▶ (Alternatively, enter the within-cells standard deviation on main screen and the between-groups standard deviation using the alternative panel.)

▶ Click *Compute f* and then click *Register f* to return to the main screen.

▶ Click *Alpha* and select the following value: *Alpha = 0.05*.

> ## Power for a test of the null hypothesis
>
> This power analysis is for a oneway fixed effects analysis of covariance with four levels. The study will include 36 cases per cell, for a total of 144 cases. The study will include a set of one covariates, which accounts for 40.0% of the variance in the dependent variable.
>
> The criterion for significance (alpha) has been set at 0.050. The analysis of variance is nondirectional (that is, two-tailed), which means that an effect in either direction will be interpreted.
>
> ## Main effects
>
> Teaching method will include four levels, with 36 cases per level. For analysis of variance, the effect size (f) is 0.250, which yields power of 0.70. For analysis of covariance, the adjusted effect size (f) is 0.32, which yields power of 0.91.

▶ On the main screen, enter the number of covariates (1) and the R-squared for covariates (0.40).

▶ Double-click the *Power adjusted for covariates* value. This tells the program to find, in the next step, the required number of cases for ANCOVA (alternatively, double-click on the power shown for ANOVA to find the required number of cases for ANOVA).

▶ Press Ctrl-F (or click the *Find N* icon if the default power has already been set to 90). The program shows that 36 cases per cell (as compared with 58 for ANOVA) will yield power of 90%.

▶ Click the *Report* icon to generate the following report:

18 Analysis of Variance/Covariance (Factorial)

Selecting the Procedure

To display the available procedures, choose *New analysis* from the File menu.

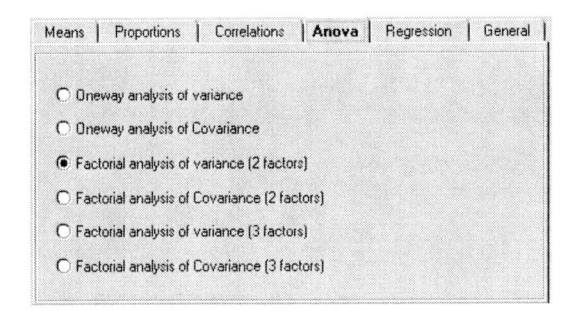

```
Means | Proportions | Correlations | Anova | Regression | General

  ○ Oneway analysis of variance
  ○ Oneway analysis of Covariance
  ● Factorial analysis of variance (2 factors)
  ○ Factorial analysis of Covariance (2 factors)
  ○ Factorial analysis of variance (3 factors)
  ○ Factorial analysis of Covariance (3 factors)
```

Application

The program can compute power for a fixed-effect balanced factorial analysis of variance with two or three factors, including interactions.

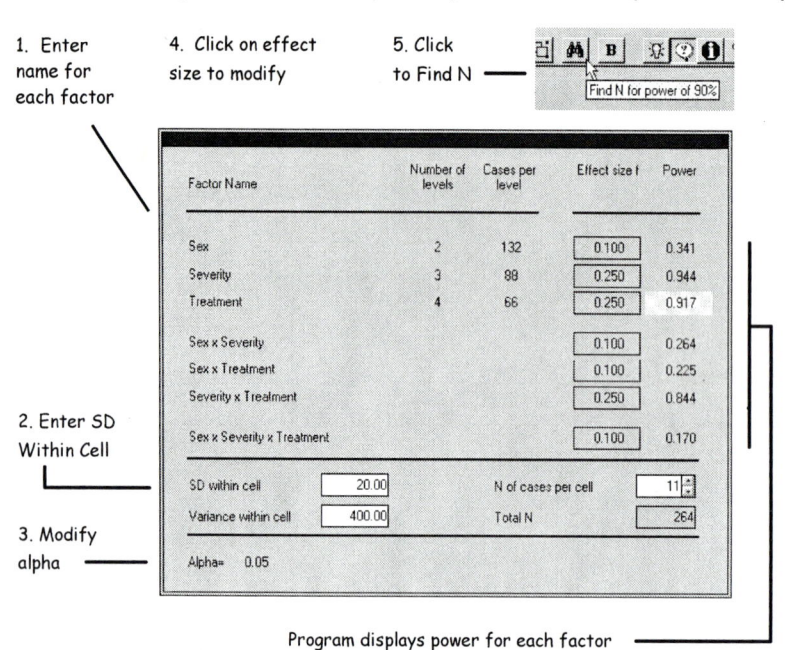

Figure 18.1 Factorial analysis of variance (three factors)

1. Enter name for each factor
2. Enter SD Within Cell
3. Modify alpha
4. Click on effect size to modify
5. Click to Find N

Factor Name	Number of levels	Cases per level	Effect size f	Power
Sex	2	132	0.100	0.341
Severity	3	88	0.250	0.944
Treatment	4	66	0.250	0.917
Sex x Severity			0.100	0.264
Sex x Treatment			0.100	0.225
Severity x Treatment			0.250	0.844
Sex x Severity x Treatment			0.100	0.170

SD within cell: 20.00
Variance within cell: 400.00
Alpha= 0.05
N of cases per cell: 11
Total N: 264

Program displays power for each factor

Effect Size

The effect size (f) used in analysis of variance is an extension of the effect size (d) used for a t-test. Recall that d is the mean difference between groups divided by the dispersion within groups. Similarly, f is based on the dispersion (standard deviation) between groups divided by the dispersion (standard deviation) within groups. (The effect size, f, is a pure measure of effect size and should not be confused with the F statistic, which takes into account sample size as well as effect size.)

In factorial ANOVA, the within-cells standard deviation is affected by the inclusion of multiple factors. For example, if the analysis includes *Gender* and *Treatment* as factors, the within-cells standard deviation is the within-gender, within-treatment standard deviation. The user should take this into account when estimating the within-cells standard deviation, which will have the effect of reducing within-cell variance and increasing the effect size.

The program allows the user to enter f for each factor directly or by using Cohen's conventions for research in the social sciences—small = 0.10, medium = 0.25, and large = 0.40—for main effects or interactions.

Alternatively, the user is allowed to provide data that the program uses to compute f. In this case, the program requires the user to provide the within-cells standard deviation

on the main screen. Then, enter data for the between-groups standard deviation in one of three formats:
- Enter the between-groups standard deviation or variance (for main effects or interactions).
- Enter the range of means for the factor (main effects only).
- Enter the mean for each group (main effects only).

The program will use the data provided in any of these formats to compute the between-groups standard deviation and then proceed to compute f.

Entering the Effect Size (f) for Factorial ANOVA

▶ On the main screen, enter the *SD within cell* value (optional if the user will enter f directly).

Figure 18.2 Factorial ANOVA (three factors) main screen

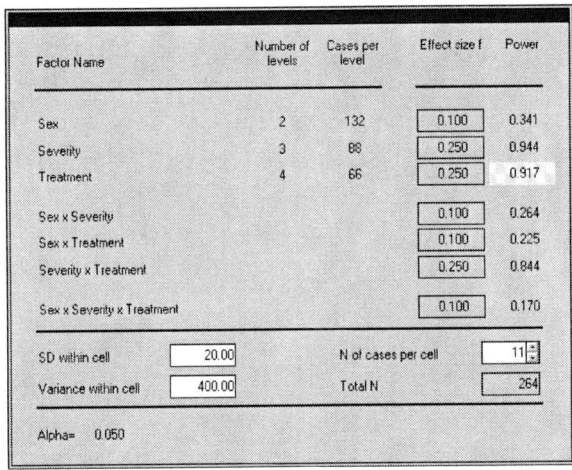

Factor Name	Number of levels	Cases per level	Effect size f	Power
Sex	2	132	0.100	0.341
Severity	3	88	0.250	0.944
Treatment	4	66	0.250	0.917
Sex x Severity			0.100	0.264
Sex x Treatment			0.100	0.225
Severity x Treatment			0.250	0.844
Sex x Severity x Treatment			0.100	0.170

SD within cell 20.00 N of cases per cell 11
Variance within cell 400.00 Total N 264

Alpha= 0.050

▶ Click on the effect size shown for any of the main effects or interactions (initially, this is 0.000). The program will immediately transfer control to one of four panels (see Figure 18.3). Choose one panel and enter the value(s) for f, standard deviation, range, or means.

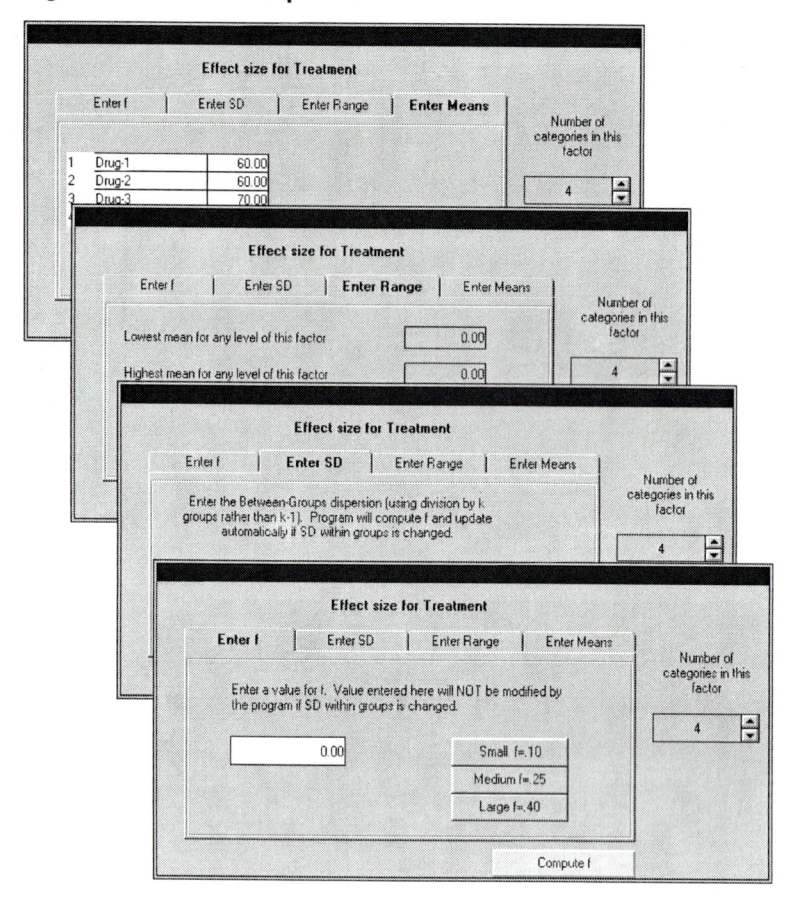

Figure 18.3 Effect size panels

▶ Repeat this process for each factor. The method used to enter the effect size is set on a factor-by-factor basis.

Analysis of Variance/Covariance (Factorial) 131

Figure 18.4 Panel to enter f directly or using conventional values

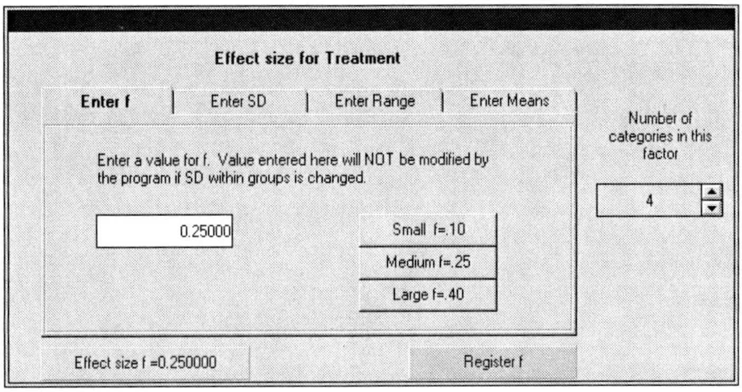

[Screenshot: Effect size for Treatment panel with tabs "Enter f", "Enter SD", "Enter Range", "Enter Means". Value 0.25000 entered. Small f=.10, Medium f=.25, Large f=.40. Number of categories in this factor: 4. Effect size f =0.250000. Register f button.]

This panel is intended for users who are familiar with the effect size, f, and feel comfortable specifying the f value directly. It is also appropriate for users who have little basis for estimating the effect size and therefore prefer to work with Cohen's conventions for small, medium, or large effects.

▶ Enter the number of groups.

▶ Click on one of the conventional values for effect size or enter a value directly. For example, a value of 0.25 would fall in the medium range according to Cohen's conventions.

▶ Click *Compute f* to compute the corresponding f.

▶ Click *Register f* to transfer the value to the main screen.

Figure 18.5 Panel to enter the between-groups standard deviation

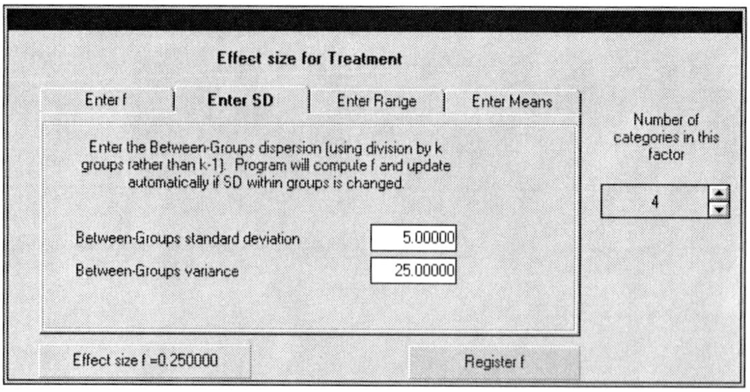

This panel is appropriate for researchers who are able to provide an estimate of the between-groups dispersion.

▶ Enter the number of groups.

▶ Enter the between-groups dispersion (either the standard deviation or the variance).

▶ Click *Compute f* to compute the corresponding f.

▶ Click *Register f* to transfer the value to the main screen.

In this example, the user entered the between-groups standard deviation (5) and the between-groups variance (25). Using this information and the within-cells standard deviation (20, entered on the main screen), the program computes the effect size, f (in this case, $5/20 = 0,25$).

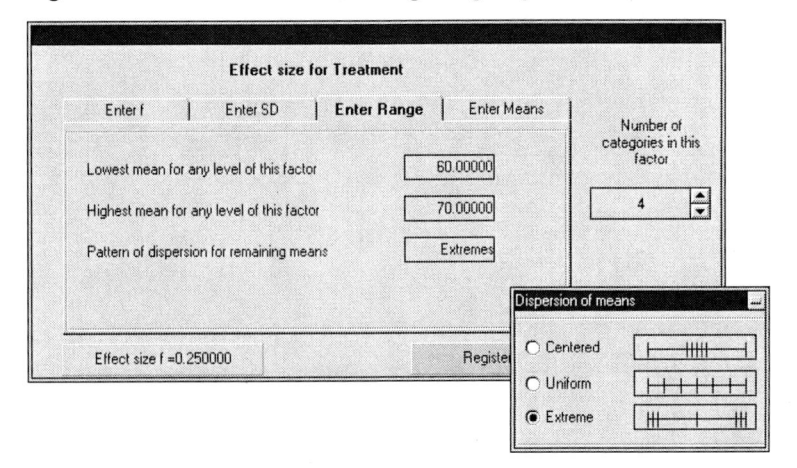

Figure 18.6 Panel to enter the range of group means

This panel is appropriate for researchers who are able to estimate the range of means but not the mean for each group.

▶ Enter the number of groups.

▶ Enter the single lowest and highest means.

▶ Once the two extreme groups have been specified, the remaining groups can fall into one of three patterns: all remaining groups fall at the center of the range (*Centered*), which will yield the smallest effect; the remaining groups are distributed evenly across the range (*Uniform*); or the remaining groups fall at either extreme (*Extreme*). Click on the value shown to activate the pop-up box and make a selection.

▶ Click *Compute f* to compute the corresponding f.

▶ Click *Register f* to transfer the value to the main screen.

Note that when the study has only two groups, the three patterns are identical and, thus, will have no effect on f.

On the main screen, the user specified that the within-groups standard deviation is 20. In Figure 18.6, the user specified that the four groups will have means that range from 60 to 70 and that the remaining two groups will have means at either extreme of this range (that is, at 60 and 70). The program has computed the corresponding f as 0.25.

Figure 18.7 Panel to enter the mean for each group

	Effect size for Treatment				
	Enter f	Enter SD	Enter Range	Enter Means	Number of categories in this factor
1	Drug-1	60.00000			
2	Drug-2	60.00000			4
3	Drug-3	70.00000			
4	Drug-4	70.00000			

Effect size f = 0.250000   Register f

This panel is appropriate for researchers who are able to provide an estimate for every one of the group means.

▶ Enter the number of groups. The program will display an entry box for each group.

▶ Enter a name (optional) and mean (required) for each group.

▶ Click *Compute f* to compute the corresponding f.

▶ Click *Register f* to transfer the value to the main screen.

In this example, the user specified mean values of 60, 60, 70, and 70. These values provide the between-groups standard deviation (5), and the user has already provided the within-cells standard deviation (20). The program computes the effect size, f, as 0.25.

Correspondence between the Four Approaches

The four methods provided for computing f are mathematically equivalent to each other. In the example, the user is working with means of 60, 60, 70, and 70. The dispersion can be described by specifying these four distinct values. It can also be described by entering the range (60 to 70) and the pattern (extreme) of the two remaining means. Finally, it can be described by entering the standard deviation for the four means (5).

Each of these approaches yields the identical value for the between-groups standard deviation (5). Given the within-cells standard deviation (20) from the main screen, all yield the identical value for f (0.25).

Of course, the different methods yield the same value only when provided with equivalent information. Often, researchers will want to estimate f using more than one method as a check that the estimate is accurate. The program will retain the data entered in any panel, but only the registered data will be transferred to the main screen.

Effect Size Updated Automatically

When you initially enter the between-groups standard deviation (either by entering the between-groups standard deviation, the range of means, or the four individual means), the program will compute f using this value and the within-cells standard deviation (which is entered on the main screen). If you later modify the within-cells standard deviation, the effect size will be adjusted automatically to reflect this change (just as the effect size, d, is modified in a t-test as you update the within-groups standard deviation).

However, if you have decided to enter the effect size by specifying f directly, the effect size is not updated. The program assumes that a user who has specified a medium effect (0.25) may not know whether the between-groups/within-groups standard deviation is 0.25/1 or 2.5/10, but merely that a medium effect would be important to detect. In this case, the value entered for the within-cells standard deviation has no impact on f.

This is handled on a factor-by-factor basis. If the effect size for *Sex* is entered directly as 0.10 while the effect size for *Treatment* is entered by providing the between-groups standard deviation, changes to the within-groups standard deviation will not affect the former but will affect the latter. All effect sizes are displayed on the main screen, and the user should check these values after making changes to the within-groups standard deviation.

Alpha

Click *Alpha* to modify this value. ANOVA is sensitive to an effect in any direction and as such is nondirectional.

Sample Size

To modify the sample size, use the spin control. By default, this will increase the number of cases by 5, but this can be modified to any number (choose *N-Cases* from the Options menu).

Click the *Find N* icon to have the program find the number of cases required for the default level of power. The default value for power is 90%, but this can be modified temporarily (Ctrl-F) or permanently (choose *Preferences* from the Options menu).

Note. When you click the *Find N* icon, the program assumes that you want to work with the first factor (that is, to find the number of cases required to yield adequate power for Factor A). To select another factor, double-click on the power shown for that factor (the selected value will be highlighted) and click the *Find N* icon again.

Example 1

A researcher plans to assign patients to one of four treatments regimens. Following a period of treatment, she will measure the patients' level of antibodies. The analysis will take into account patient gender (two levels), disease severity at baseline (three levels), and treatment (four levels). The researcher is primarily interested in the effect of treatment and in the treatment-by-severity interaction. A clinically important effect for either would be a medium effect of f = 0.25.

▶ Choose *Factorial analysis of variance (3 factors)*.

▶ Enter a name for each factor—*Sex*, *Severity*, and *Treatment*. Names for the interactions are assigned automatically.

▶ Click on the effect size for the first factor (*Sex*). Specify two levels with f = 0.10.

▶ Click on the effect size for the second factor (*Severity*). Specify three levels with f = 0.25.

▶ Click on the effect size for the third factor (*Treatment*). Specify four levels with f = 0.25.

▶ Click on the effect size for each of the interactions. In this example, the effect size is set at 0.25 for the interaction of *Severity* by *Treatment* and at 0.10 for all other interactions.

▶ Click *Alpha* and select the following value: *Alpha = 0.05*.

▶ Double-click on the power shown for *Treatment*. Then press Ctrl-F (or click the *Find N* icon if the default power has already been set to 90). The program shows that 11 cases per cell (the total number of cases is 264) will yield power of 92% for *Treatment*, which is the designated factor. The power to find an effect for the interaction of *Treatment* by *Severity* is 84%.

▶ Double-click on the power shown for the interaction of *Severity* by *Treatment*. Then click the *Find N* icon. The program shows that 13 cases per cell (the total number of cases is 312) will yield power of 91% for this interaction. With this sample size, the power to find an effect for *Treatment* is 96%.

▶ Click the *Report* icon to generate the following report:

Power for a test of the null hypothesis

This power analysis is for a 2 x 3 x 4 fixed-effects analysis of variance. The study will include 13 cases per cell in a balanced design, for a total of 312 cases.

The criterion for significance (alpha) has been set at 0.05. The analysis of variance is nondirectional, which means that an effect in either direction will be interpreted.

Main effects

Sex will include two levels, with 156 cases per level. The effect size (f) is 0.100, which yields power of 0.397.

Severity will include three levels, with 104 cases per level. The effect size (f) is 0.250, which yields power of 0.974.

Treatment will include four levels, with 78 cases per level. The effect size (f) is 0.250, which yields power of 0.959.

Interactions

Sex x Severity. The effect size (f) is 0.100, which yields power of 0.310.

Sex x Treatment. The effect size (f) is 0.100, which yields power of 0.265.

Severity x Treatment. The effect size (f) is 0.250, which yields power of 0.911.

Sex x Severity x Treatment. The effect size (f) is 0.100, which yields power of 0.199.

Factorial Analysis of Covariance

Analysis of covariance is identical to analysis of variance except for the presence of the covariate. The covariates are able to explain some of the variance in the outcome measure. This serves to reduce the error term, which yields a more powerful test.

We will illustrate the use of this procedure by extending the previous example. As in Example 1, patients are assigned at random to one of four treatment groups, treated for a period of time, and then assessed on a continuous measure. In the ANOVA example, we simply assessed the level of antibodies at the end point. For the ANCOVA example, we will assume that the patients are assessed at the baseline as well and that the baseline ratings serve as a covariate.

The specification of effect size by the user proceeds exactly as that for ANOVA. Specifically, the within-groups standard deviation and the between-groups standard deviation are entered as though no covariate is present. If the user chooses to enter f directly, enter f as though no covariate is present. The covariate serves to reduce the error term (which boosts the effect size, f), but this is taken into account by the program, which displays both the unadjusted f and the adjusted f.

Figure 18.8 shows the factorial analysis of covariance. We assume that the baseline scores account for 40% of the scores at the end point.

Analysis of Variance/Covariance (Factorial)   139

Figure 18.8   Factorial analysis of covariance (three factors)

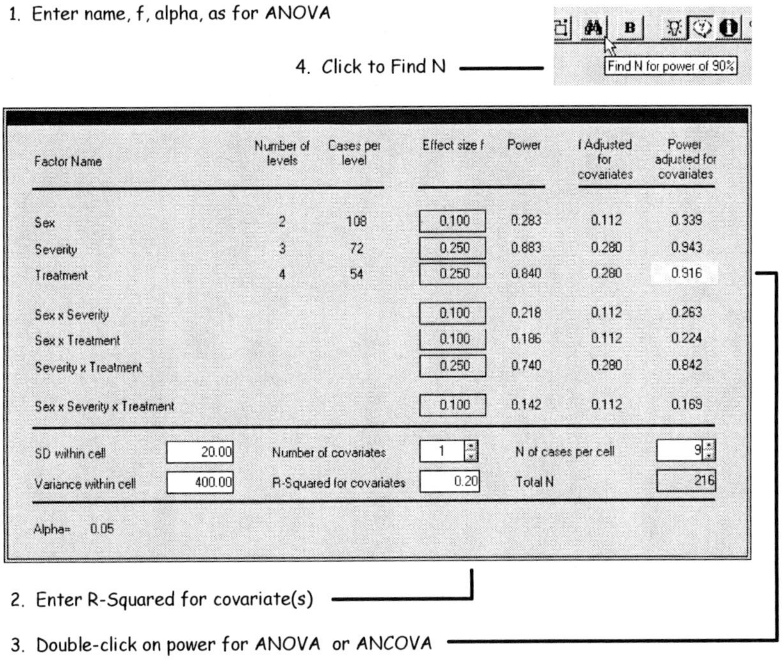

1. Enter name, f, alpha, as for ANOVA

4. Click to Find N

Factor Name	Number of levels	Cases per level	Effect size f	Power	f Adjusted for covariates	Power adjusted for covariates
Sex	2	108	0.100	0.283	0.112	0.339
Severity	3	72	0.250	0.883	0.280	0.943
Treatment	4	54	0.250	0.840	0.280	0.916
Sex x Severity			0.100	0.218	0.112	0.263
Sex x Treatment			0.100	0.186	0.112	0.224
Severity x Treatment			0.250	0.740	0.280	0.842
Sex x Severity x Treatment			0.100	0.142	0.112	0.169

SD within cell 20.00   Number of covariates 1   N of cases per cell 9
Variance within cell 400.00   R-Squared for covariates 0.20   Total N 216

Alpha= 0.05

2. Enter R-Squared for covariate(s)

3. Double-click on power for ANOVA or ANCOVA

Example 2

A researcher plans to assign patients to one of four treatments regimens. Following a period of treatment, she will measure the patients' level of antibodies. The analysis will take into account patient gender (two levels), disease severity at baseline (three levels), and treatment (four levels). The researcher is primarily interested in the effect of treatment and in the treatment-by-severity interaction. A clinically important effect for either would be a medium effect of f = 0.25. The pre-score will serve as a covariate and is expected to account for 20% of the variance in post-scores.

▶ Choose *Factorial analysis of covariance (3 factors)*.

▶ Enter a name for each factor—*Sex*, *Severity*, and *Treatment*. Names for the interactions are assigned automatically.

▶ Click on the effect size for the first factor (*Sex*). Specify two levels with f = 0.10.

▶ Click on the effect size for the second factor (*Severity*). Specify three levels with f = 0.25.

▶ Click on the effect size for the third factor (*Treatment*). Specify four levels with f = 0.25.

▶ Click on the effect size for each of the interactions. In this example, the effect size is set at 0.25 for the interaction of *Severity* by *Treatment* and at 0.10 for all other interactions.

▶ Click *Alpha* and select the following value: *Alpha = 0.05*.

▶ On the main screen, enter the number of covariates (1) and the *R*-squared for covariates (0.20).

▶ Double-click the *Power adjusted for covariates* value in the *Treatment* row. Press Ctrl-F (or click the *Find N* icon if the default power has already been set to 90). The program shows that 9 cases per cell (the total number of cases is 216) will yield a power of 92% for *Treatment*, which is the designated factor. The power to find an effect for the interaction of *Treatment* by *Severity* is 84%.

▶ Double-click on the power shown for the interaction of *Severity* by *Treatment*. Then click the *Find N* icon. The program shows that 11 cases per cell (the total number of cases is 264) will yield a power of 92% for this interaction. With this sample size, power to find an effect for *Treatment* is 97%.

▶ Click the *Report* icon to generate the following report:

Power for a test of the null hypothesis

This power analysis is for a 2 x 3 x 4 fixed effects analysis of covariance. The study will include 11 cases per cell in a balanced design, for a total of 264 cases. The study will include a set of one covariate, which accounts for 20.0% of the variance in the dependent variable.

The criterion for significance (alpha) has been set at 0.05. The analysis of variance is nondirectional, which means that an effect in either direction will be interpreted.

Main effects

Sex will include two levels, with 132 cases per level. For analysis of variance, the effect size (f) is 0.100, which yields power of 0.341. For analysis of covariance, the adjusted effect size (f) is 0.112, which yields power of 0.409.

Severity will include three levels, with 88 cases per level. For analysis of variance, the effect size (f) is 0.250, which yields power of 0.944. For analysis of covariance, the adjusted effect size (f) is 0.280, which yields power of 0.979.

Treatment will include four levels, with 66 cases per level. For analysis of variance, the effect size (f) is 0.250, which yields power of 0.917. For analysis of covariance, the adjusted effect size (f) is 0.280, which yields power of 0.965.

Interactions

Sex x Severity. For analysis of variance, the effect size (f) is 0.100, which yields power of 0.264. For analysis of covariance, the adjusted effect size (f) is 0.112, which yields power of 0.320.

Sex x Treatment. For analysis of variance, the effect size (f) is 0.100, which yields power of 0.225. For analysis of covariance, the adjusted effect size (f) is 0.112, which yields power of 0.274.

Severity x Treatment. For analysis of variance, the effect size (f) is 0.250, which yields power of 0.844. For analysis of covariance, the adjusted effect size (f) is 0.280, which yields power of 0.923.

Sex x Severity x Treatment. For analysis of variance, the effect size (f) is 0.100, which yields power of 0.170. For analysis of covariance, the adjusted effect size (f) is 0.112, which yields power of 0.205.

Generate Table

To generate a table, click the *Table* icon. For a two-factor ANOVA, the table will show power for Factor A, Factor B, and the interaction as a function of sample size. For a three-factor ANOVA, the table will show power for Factor A, Factor B, and Factor C as a function of effect size. The table will not include power for interactions. In either case, the program can include alpha as a second factor as well.

Generate Graph

To generate a graph, click the *Graph* icon. For a two-factor ANOVA, the program will graph power for Factor A, Factor B, and the interaction as a function of sample size. For a three-factor ANOVA, the program will graph power for Factor A, Factor B, and Factor C as a function of sample size. In either case, the program will graph power as a function of alpha for Factor A only.

19 Multiple Regression

Selecting the Procedure

To display the available procedures, choose *New analysis* from the File menu.

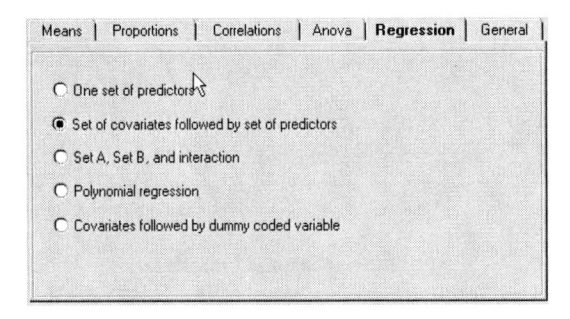

- One set of predictors
- ● Set of covariates followed by set of predictors
- Set A, Set B, and interaction
- Polynomial regression
- Covariates followed by dummy coded variable

Application

Figure 19.1 Set A, set B, and interaction

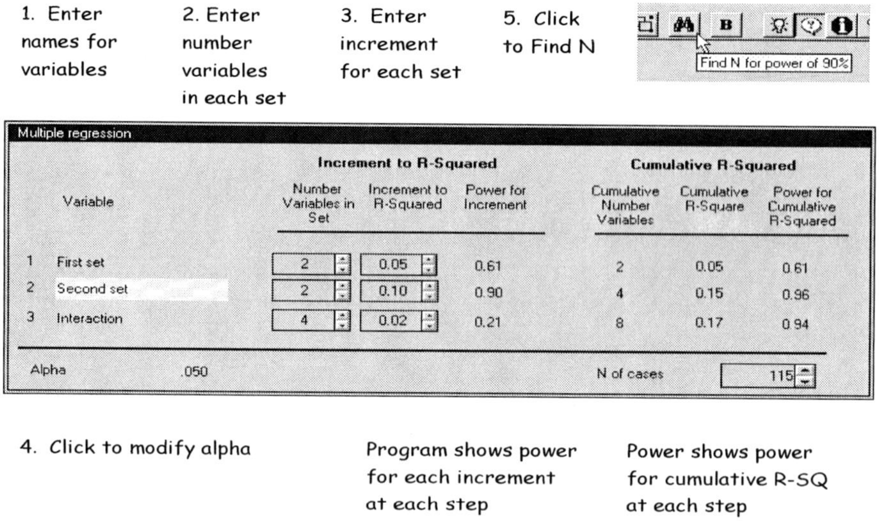
1. Enter names for variables
2. Enter number variables in each set
3. Enter increment for each set
5. Click to Find N
Find N for power of 90%
Multiple regression
Increment to R-Squared
Cumulative R-Squared
Variable
Number Variables in Set
Increment to R-Squared
Power for Increment
Cumulative Number Variables
Cumulative R-Square
Power for Cumulative R-Squared
1 First set 2 0.05 0.61 2 0.05 0.61
2 Second set 2 0.10 0.90 4 0.15 0.96
3 Interaction 4 0.02 0.21 8 0.17 0.94
Alpha .050
N of cases 115
4. Click to modify alpha
Program shows power for each increment at each step
Power shows power for cumulative R-SQ at each step

Multiple regression is used to study the relationship between sets of independent (predictor) variables and a single dependent variable. The "set" of variables is defined broadly and may consist of a single variable, such as age, or a set of variables that together serve as an index of some construct (for example, disease severity and duration together serve as an indicator of prognosis).

The variables may be continuous, in which case the actual rating (or some transformation thereof) is used as the predictor. The variables may also be categorical, in which case dummy coding (or one of its variants) is used to create a set of independent variables (sex is coded as 0 or 1).

Typically, sets of variables are entered into the multiple regression in a predetermined sequence. At each point in the analysis, the researcher may test the significance of the increment (the increase in R^2 for the new set over and above previous sets) or the significance of all variables in the equation. The program is able to compute power for either of these tests.

- At left, each set is identified by name
- The panel of columns labeled *Increment to R-Squared* shows data for the increment due to each set—the user provides the number of variables and the increment, and the program computes power.
- The panel of columns to the right shows data for the cumulative R^2. All data here (the cumulative number of variables, the cumulative R^2, and power) are computed automatically.

The Designated Set

The program allows the user to designate one set as the set of primary interest (choose *Customize screen* from the Options menu, and then select *Display superset.*)

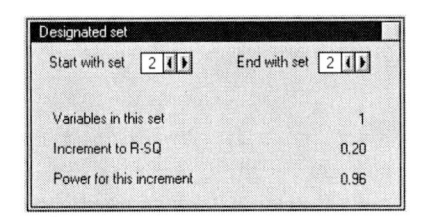

Designated set	
Start with set 2	End with set 2
Variables in this set	1
Increment to R-SQ	0.20
Power for this increment	0.96

- This set may be identical with a set in the main panel, or it may include two or more contiguous sets.
- The program will display power for the increment attributed to this set.
- The program's tools (Find N, Report, Table, and Graph) will provide data for this set.

Effect Size

Effect size for multiple regression is given by f^2, defined as explained variance/error variance. This is similar to the index (f) used for ANOVA, except that f is based on the standard deviations, while f^2 is based on the variances, as is common in multiple regression. Cohen (1988) notes that f^2 can be thought of as a kind of signal-to-noise ratio.

When used for a single set of variables, f^2 is equal to $R^2/(1-R^2)$. When used for more than one set of variables, R^2 in the numerator is the increment to R^2 for the current set, while R^2 in the denominator is the cumulative R^2 for all sets in the regression (but see "Error Model" on p. 148).

Cohen provides the following conventions for research in the social sciences: small ($f^2 = 0,02$), medium ($f^2 = 0,15$) and large ($f^2 = 0,35$). Assuming a single set of variables, these would correspond to R^2 values of about 0.02, 0.13, and 0.26.

Alpha and Tails

Click *Alpha* to modify it. Multiple regression is sensitive to effects in any direction and thus is naturally two-tailed.

Sample Size

To modify the sample size, click the spin control for *N of cases*. To find the number of cases required to yield the required power, press Ctrl-F or click the *Find N* icon. The program will find the number of cases required to yield required power for the set defined in the Designated set box.

Computational Options for Power

Error Model

Multiple regression can be run with the model 1 error or the model 2 error (not to be confused with the type 1 error versus the type 2 error, or with one-tail versus two-tail). Choose *Computational formulas* from the Options menu.

[Error Model dialog: Model 1 error / ● Model 2 error]

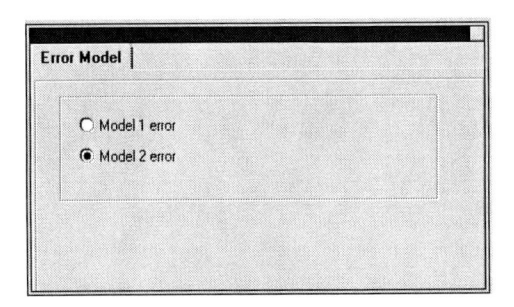

The two models differ in the definition of the error term used to compute the effect for each element in the model. In either model, F^2 is defined as R^2/Error. In model 1, error is defined as $1 - R^2$ through the current set. In model 2, error is defined as $1 - R^2$ for all variables in the model.

Assume, for example, that we will enter a set of covariates (*Baseline rating*), followed by a main set. We want to compute power for the covariates. The data are shown here.

	Variable	Number Variables in Set	Increment to R-Squared	Power for Increment	Cumulative Number Variables	Cumulative R-Square	Power for Cumulative R-Squared
1	Baseline rating	1	0.40	1.00	1	0.40	1.00
2	Treatment	1	0.20	0.96	2	0.60	1.00

Alpha .010   N of cases 41

With only the covariates in the equation, R^2 is 0.40, meaning that $(1 - 0,40)$, or 60% of the variance, is classified as error.

With both sets of variables in the model, the cumulative R^2 is 0.60, which means that the unexplained variance is $(1 - 0,60)$, or 40%.

The effect size for the covariates would be computed as $0,40/0,60 = 0,66$ under model 1, or as $0,40/0,40 = 1,00$ under model 2.

The effect size for set B is $0,20/0,40 = 0,50$ under either model, since set B is entered at the last step and all other variables are already in the equation. If there were additional sets to be entered subsequently to set B, then the selection of an error model would affect set B as well.

The default in this program (as in most statistical analysis programs) is the model 2 error. This yields higher power in most cases, but if the number of cases is low (so that small changes in the degrees of freedom are important) and/or subsequent sets incorporate many variables (thus consuming degrees of freedom), while yielding a small increment, it is possible that model 1 could yield a more powerful test.

Options for Study Design

Templates

The program comes with templates corresponding to some of the more common types of analyses (choose *Data entry/Study design* from the Options menu).

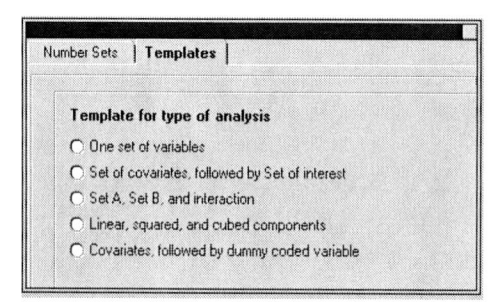

```
Number Sets | Templates

Template for type of analysis
  ○ One set of variables
  ○ Set of covariates, followed by Set of interest
  ○ Set A, Set B, and interaction
  ○ Linear, squared, and cubed components
  ○ Covariates, followed by dummy coded variable
```

Customizing the Study Design

The templates cover some typical situations but are intended only as a starting point. The program allows the user to enter any number of sets (1 – 10), and each set can include as many as 99 variables (choose *Data entry/Study design* from the Options menu).

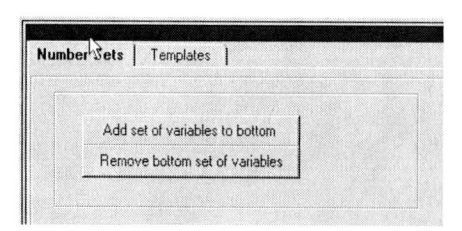

```
Number Sets | Templates

  Add set of variables to bottom
  Remove bottom set of variables
```

Generating a Table

Click the *Table* icon to generate a table of power by sample size. The table in multiple regression will show power for only one effect size (the designated effect). The table can be modified to show power as a function of alpha as well as sample size.

Generating a Graph

Click the *Graph* icon to generate a graph of power by sample size.
The graph in multiple regression will display only one effect size (the designated effect) but can be modified to show power as a function of sample size and alpha.

Example 1

One Set of Variables

A researcher believes that aptitude, which she has operationally defined as scores on four tests, will predict students' grade point average (GPA). A meaningful effect is defined as one that explains 10% of the variance in the GPA. The study will be run with alpha at 0.05.

▶ On the Regression panel, select *One set of predictors*. (Or, within the module, choose *Data entry/Study design* from the Options menu, click the *Templates* tab, and select the template *One set of variables*.)

▶ Enter the number of variables—4.

▶ Enter the *Increment to R-Squared*—0.10.

▶ Click *Alpha* and select the value 0.05.

▶ Press Ctrl-F (or click the *Find N* icon if the default power has already been set to 80). The program shows that a sample of 113 cases will yield power of 80% for the given parameters.

Find N required for power of 80%

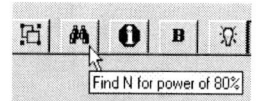

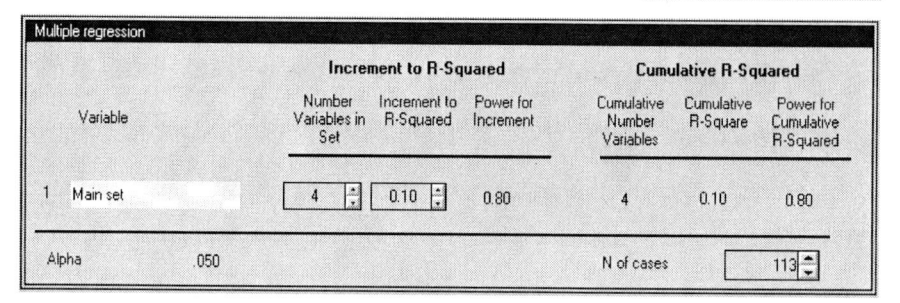

Multiple regression		Increment to R-Squared			Cumulative R-Squared		
	Variable	Number Variables in Set	Increment to R-Squared	Power for Increment	Cumulative Number Variables	Cumulative R-Square	Power for Cumulative R-Squared
1	Main set	4	0.10	0.80	4	0.10	0.80

Alpha .050   N of cases 113

▶ Click the *Report* icon to generate the following report:

Power for a test of the null hypothesis

The model will include (A) zero covariates, which will yield an R-squared of 0.000. It will include (B) four variables in the set of interest, which will yield an increment of 0.100. The model will also include (C) zero variables entered subsequently to the set of interest, which account for an additional 0.000 of variance. The total R-squared for the four variables in the model is 0.100.

The power analysis focuses on the increment for the set of interest (B) over and above any prior variables (that is, four variables yielding an increment of 0.10). With the given sample size of 113 and alpha set at 0.05, the study will have power of 0.80

The test is based on the model 2 error, which means that variables entered into the regression subsequently to the set of interest will serve to reduce the error term in the significance test and are therefore included in the power analysis.

This effect was selected as the smallest effect that would be important to detect, in the sense that any smaller effect would not be of clinical or substantive significance. It is also assumed that this effect size is reasonable, in the sense that an effect of this magnitude could be anticipated in this field of research.

Example 2

Set of Covariates Followed by Set of Interest

Patients are being recruited for a clinical trial that will assess the impact of a drug on asthma symptoms. Patients' symptoms at baseline will serve as a covariate. Patients' drug assignments (dummy coded as 0 for placebo and 1 for drug) will serve as the variable of interest. Symptom level following treatment will serve as the dependent variable.

It is expected that the baseline rating will explain 40% of the variance in outcome. The treatment would be clinically useful if it could explain an additional 20% of the variance. This will serve as the effect size in the power analysis.

The study will be run with alpha set at 0.01, and the researcher wants to compute the sample size required for power of 95%.

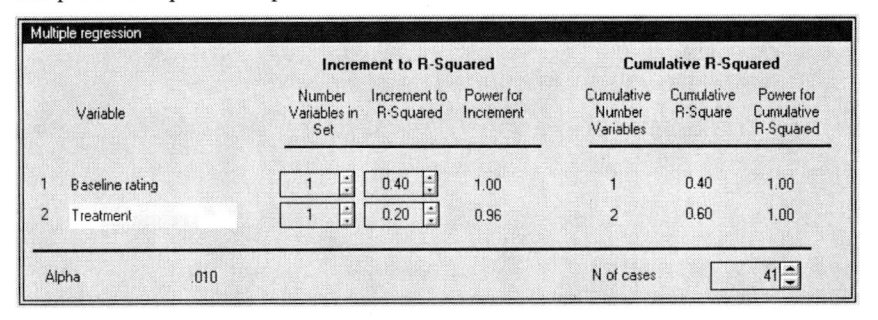

	Variable	Number Variables in Set	Increment to R-Squared	Power for Increment	Cumulative Number Variables	Cumulative R-Square	Power for Cumulative R-Squared
1	Baseline rating	1	0.40	1.00	1	0.40	1.00
2	Treatment	1	0.20	0.96	2	0.60	1.00

Alpha .010   N of cases 41

▶ On the Regression panel, select *Set of covariates followed by set of predictors*. (Or, within the module, choose *Data entry/Study design* from the Options menu, click the *Templates* tab, and select the template *Set of covariates followed by set of interest*.)

▶ On line 1, enter *Baseline rating*, with one variable and 40% of the variance explained.

▶ On line 2, enter *Treatment*, with one variable and 20% of the variance explained.

▶ Click *Alpha* and select the value 0.01.

The next step is to find the number of cases required for power of 99%. However, the program is now showing data for several analyses—power for the covariates, power for the increment due to treatment, and power for the covariates and treatment combined. Therefore, we need to identify the test for which we want 99% power.

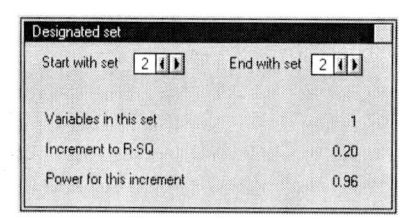

> **Designated set**
> Start with set [2] End with set [2]
>
> Variables in this set   1
> Increment to R-SQ   0.20
> Power for this increment   0.96

▶ The program displays a box for this designated set, as shown above. (If the Designated set box is not displayed, choose *Customize screen* from the Options menu and select *Display superset*.) Using the spin controls, define the set as including set 2 only.

- On the main panel, set 2 (*Treatment*) is now highlighted.
- In line 2, the number of variables is shown as 1.
- In line 2, the increment is shown as 20%.

The program will focus on this designated set when it finds the number of cases required for power.

▶ Press Ctrl-F and click 0.95. The program shows that a sample of 41 cases will yield power of 96% for the designated set (that is, the increment of treatment over the covariates).

Note: The *N of cases* shown is 41 (power = 0.96). If the number of cases is set manually to 40, power will display as 0.95, which suggests that this would have met the criterion. However, if additional digits are displaced (choose *Decimals* from the Options menu), the program displays power for N = 40 as 0.949 rather than 0.95.

▶ Click the *Report* icon to generate the following report:

> **Power for a test of the null hypothesis**
>
> The model will include (A) one covariate, which will yield an R-squared of 0.400. It will include (B) one variable in the set of interest, which will yield an increment of 0.200. The model will also include (C) zero variables entered subsequently to the set of interest, which accounts for an additional 0.000 of variance. The total R-squared for the two variables in the model is 0.600.
>
> The power analysis focuses on the increment for the set of interest (B) over and above any prior variables (that is, one variable yielding an increment of 0.20). With the given sample size of 41 and alpha set at 0.001, the study will have power of 0.96
>
> The test is based on the model 2 error, which means that variables entered into the regression subsequently to the set of interest will serve to reduce the error term in the significance test and are therefore included in the power analysis.
>
> This effect was selected as the smallest effect that would be important to detect, in the sense that any smaller effect would not be of clinical or substantive significance. It is also assumed that this effect size is reasonable, in the sense that an effect of this magnitude could be anticipated in this field of research.

Example 3

Two Sets of Variables and Their Interaction

Assume that we want to study the impact of two drugs on the white blood cell count (WBC). We anticipate that the use of either drug will reduce the WBC by a modest amount, but the use of both drugs will have a synergistic effect, and we want to test the impact of this interaction. The smallest effect that would be important to detect would be an increment of 10%. The study will be run with alpha of 0.01, and we want to find the number of cases required for power of 90%

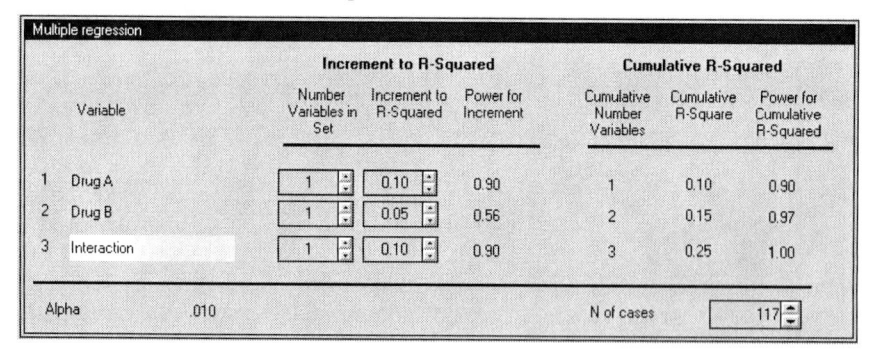

Variable	Number Variables in Set	Increment to R-Squared	Power for Increment	Cumulative Number Variables	Cumulative R-Square	Power for Cumulative R-Squared
1 Drug A	1	0.10	0.90	1	0.10	0.90
2 Drug B	1	0.05	0.56	2	0.15	0.97
3 Interaction	1	0.10	0.90	3	0.25	1.00

Alpha .010   N of cases 117

▶ On the Regression panel, select *Set A, Set B, and interaction*. (Or, within the module, choose *Data entry/Study design* from the Options menu, click the *Templates* tab, and select that template.

▶ For line 1, enter *Drug A*, with one variable and 10% of the variance explained.

▶ For line 2, enter *Drug B*, with one variable and 5% of the variance explained.

▶ For line 3 (*Interaction*), enter one variable and 10% of the variance explained.

▶ Click *Alpha* and select the value 0.01.

The program will display power for each test on the screen, but we need to identify one test that will be addressed by the report, tables, graphs, and the Find N procedure.

▶ If the Designated set box is not displayed, choose *Customize screen* from the Options menu, and select *Display superset*. Using the spin controls, define the set as including set 3 only. The box shows the designated set as beginning and ending with Set 3 (the interaction), which incorporates one variable and an increment of 10%.

▶ Press Ctrl-F (or click the *Find N* icon if the default power has already been set to 90). The program shows that a sample of 117 cases will yield power of 90% for the test just cited.

Optionally, click the *Report*, *Table*, or *Graph* icon to generate additional data, Any of these features will work with the set designated in the box.

Example 4

Polynomial Regression

Patients being treated with neuroleptics will be followed for three months and then assessed for signs of emerging abnormal movements (rated on a five-point scale). The study will include patients in the age range of 20–80. It is anticipated that the risk of these movements increases with age, but the risk is not linear. We anticipate that the risk rises for patients over the age of 50 and rises again, sharply, for patients in their 70's. We plan to enter as variables *age*, *age-squared*, and *age-cubed*.

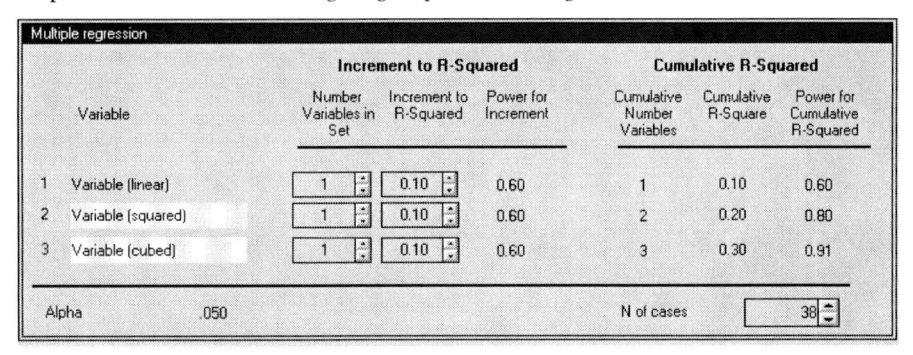

	Variable	Number Variables in Set	Increment to R-Squared	Power for Increment	Cumulative Number Variables	Cumulative R-Square	Power for Cumulative R-Squared
1	Variable (linear)	1	0.10	0.60	1	0.10	0.60
2	Variable (squared)	1	0.10	0.60	2	0.20	0.80
3	Variable (cubed)	1	0.10	0.60	3	0.30	0.91

Alpha .050   N of cases 38

▶ On the Regression panel, select *Polynomial regression*. (Or, within the module, choose *Data entry/Study design* from the Options menu, click the *Templates* tab, and select the template *Linear, squared, and cubed components*.

▶ For line 1, enter *Age*, with one variable and 10% of the variance explained.

▶ For line 2, enter *SQ*, with one variable and 10% of the variance explained.

▶ For line 3, enter *CU*, with two variables and 10% of the variance explained.

▶ Click *Alpha* and select the value 0.05.

To test the hypothesis that the relationship between age and outcome is nonlinear, the researcher wants to treat *age-squared* and *age-cubed* as a set and test the increment of this set over (linear) *age*.

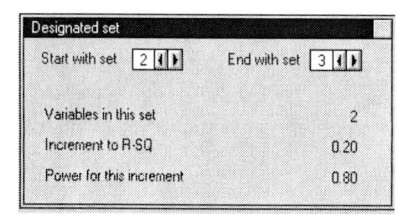

▶ If the Designated set box is not displayed, choose *Customize screen* from the Options menu and select *Display superset*. Using the spin controls, define the set as including lines 2 and 3. In the main panel, both lines are highlighted, and the set is defined as having two variables and an increment of 20%.

▶ Press Ctrl-F (or click the *Find N* icon if the default power has already been set to 80). The program shows that a sample of 38 cases will yield power of 80% for the two-variable set.

Alternatively, you could set the main panel to have only two lines, with *age-squared* and *age-cubed* combined on line 2. The number of variables for line 2 would be 2, with 20% of the variance explained.

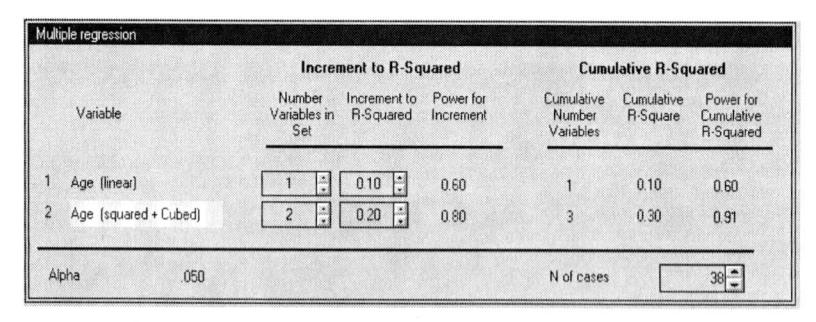

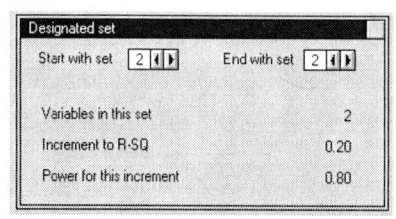

As shown here, either approach will yield the same result. Putting the three variables on three lines, however, allows the researcher to quickly look at power for any combination of sets (the increment for *age-squared*, the increment for *age-cubed*, the increment for *age-squared* combined with *age-cubed*, and so on) without the need to reenter data.

Example 5

One Set of Covariates, Two Sets of Variables, and Interactions

A hospital is planning a study to assess treatment strategies for schizophrenia. The dependent variable is the caregiver's evaluation of the patient's overall functioning at the end of a year. A similar index at study entry will serve as a covariate. The researcher wants to test the impact of the drug assignment, the incremental impact of the support, and the hypothesis that the entire treatment program is having an effect.

▶ On the Regression panel, select *Set of covariates followed by set of predictors*. (Or, within the module, choose *Data entry/Study design* from the Options menu, click the *Templates* tab, and select the template *Set of covariates followed by set of interest*.)

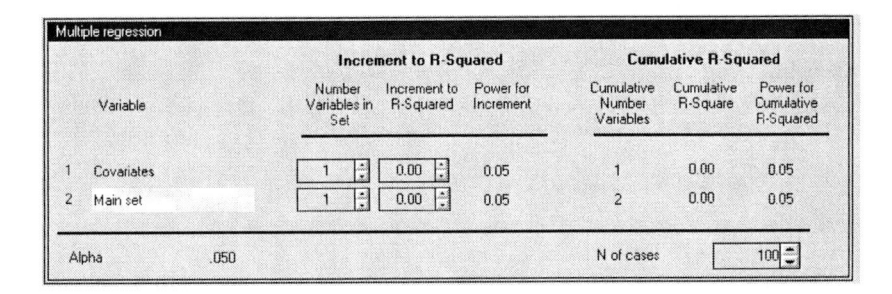

	Variable	Number Variables in Set	Increment to R-Squared	Power for Increment	Cumulative Number Variables	Cumulative R-Square	Power for Cumulative R-Squared
1	Covariates	1	0.00	0.05	1	0.00	0.05
2	Main set	1	0.00	0.05	2	0.00	0.05

Alpha .050   N of cases 100

▶ Add an additional row (choose *Data Entry/Study design* from the Options menu).

Number Sets | Templates

Add set of variables to bottom
Remove bottom set of variables

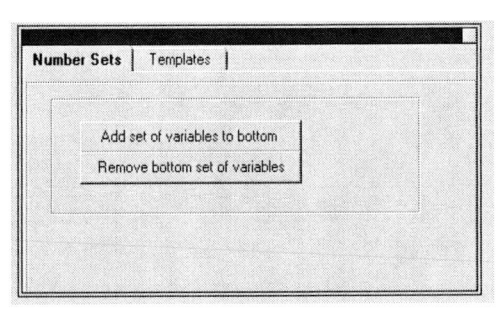

▶ For line 1, enter *Baseline ratings*, with one variable and 20% of the variance explained.

▶ For line 2, enter *Drug Group*, with one variable and 10% of the variance explained.

▶ For line 3, enter *Support Group*, with two variables and 10% of the variance explained.

	Variable	Number Variables in Set	Increment to R-Squared	Power for Increment	Cumulative Number Variables	Cumulative R-Square	Power for Cumulative R-Squared
1	Baseline ratings	1	0.20	0.99	1	0.20	0.99
2	Drug Group	1	0.10	0.87	2	0.30	1.00
3	Support Group	2	0.10	0.80	4	0.40	1.00

Alpha .050   N of cases 63

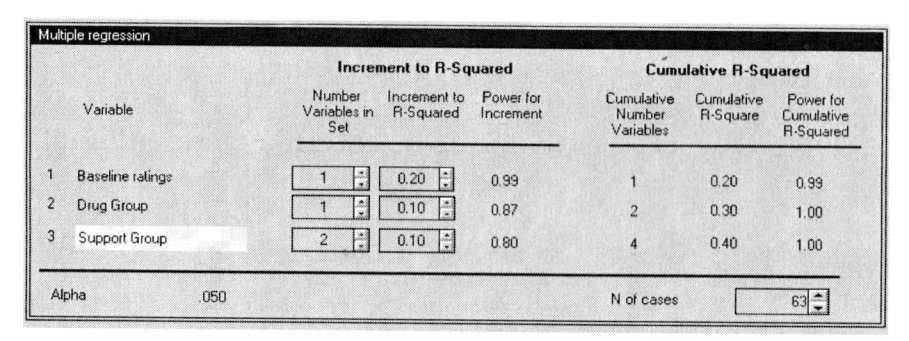

The program shows immediately that power is lower for the support group than for the drug group (for each, we are using 10% as the proportion of variance explained, but the support group requires two variables in order to yield this level of prediction). To ensure that power will be adequate for both factors, we must ensure adequate power for the support group, knowing that power for the drug effect will be higher.

▶ In the main panel, identify *Support Group* as the designated set. If the Designated set box is not displayed, choose *Customize screen* from the Options menu and select *Display superset*. Using the spin controls, define the set as beginning and ending with set 3.

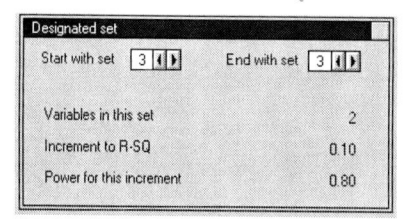

Designated set	
Start with set 3 End with set 3	
Variables in this set	2
Increment to R-SQ	0.10
Power for this increment	0.80

▶ Press Ctrl-F (or click the *Find N* icon if the default power has already been set to 80). The program shows that 63 cases will yield power of 80% for the test of the support group (the power is displayed in the Designated set box and also on the main panel, in line 3).

The main panel also shows that power for the drug group effect is 0.87, and that power for the cumulative set of baseline, drug group and support group approaches 1.00.

Note that power for the entire treatment plan (drug and support, as an increment over baseline) is not displayed on the main panel. However, it can be displayed in the Designated set box.

▶ In the Designated set box, use the spin controls to define the set as including sets 2 and 3. These sets are highlighted on the main panel. The Designated set box shows that power for these combined sets is 0.97.

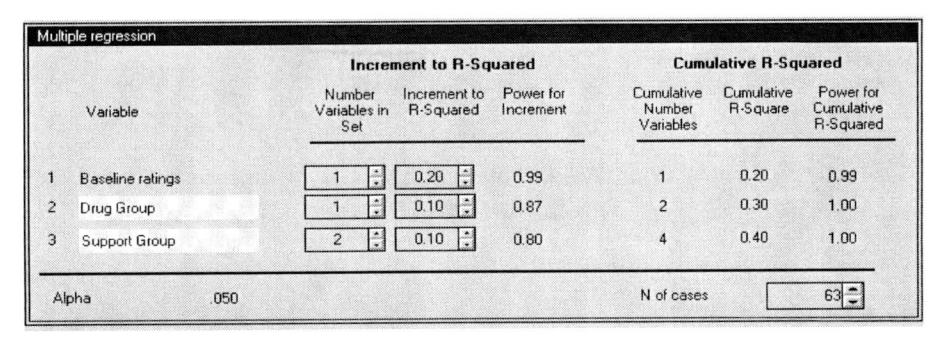

	Variable	Number Variables in Set	Increment to R-Squared	Power for Increment	Cumulative Number Variables	Cumulative R-Square	Power for Cumulative R-Squared
1	Baseline ratings	1	0.20	0.99	1	0.20	0.99
2	Drug Group	1	0.10	0.87	2	0.30	1.00
3	Support Group	2	0.10	0.80	4	0.40	1.00

Alpha .050   N of cases 63

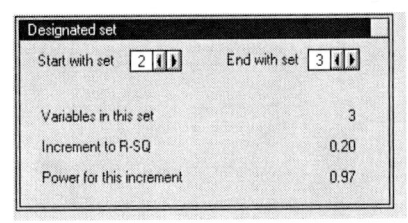

Designated set

Start with set 2   End with set 3

Variables in this set   3
Increment to R-SQ   0.20
Power for this increment   0.97

Power for the Combination of All Three Sets

You can also define the designated set as including all three sets as shown below. Power for the designated set is shown as 0.9997 (with four decimals displayed for the purpose of this example), which is the same as the value shown in the main panel for the cumulative power of the three sets.

Multiple regression

	Variable	Increment to R-Squared			Cumulative R-Squared		
		Number Variables in Set	Increment to R-Squared	Power for Increment	Cumulative Number Variables	Cumulative R-Square	Power for Cumulative R-Squared
1	Baseline ratings	1	0.20	0.9926	1	0.20	0.9926
2	Drug Group	1	0.10	0.8748	2	0.30	0.9989
3	Support Group	2	0.10	0.8007	4	0.40	0.9997

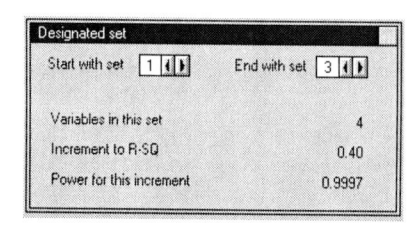

Alpha .050   N of cases 63

Designated set

Start with set 1   End with set 3

Variables in this set	4
Increment to R-SQ	0.40
Power for this increment	0.9997

20 General Case

By selecting the General panel, the advanced user can compute power for procedures not otherwise included in the program.

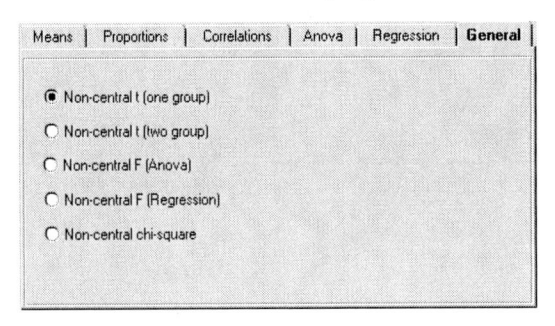

- Non-central t (one group)
- Non-central t (two group)
- Non-central F (Anova)
- Non-central F (Regression)
- Non-central chi-square

General Case (Non-Central T)

Figure 20.1 Non-central t (one group)

```
┌─────────────────────────────────────┐
│           Non-central t             │
├─────────────────────────────────────┤
│  Alpha                 0.05000000   │
│  Tails                          2   │
│                                     │
│  df numerator                   1   │
│  df denominator               198   │
│  t required for significance  1.97201748 │
│  Non-Centrality parameter  3.53553391 │
│                                     │
│  Power in expected direction  0.94042718 │
│  Power in reverse direction   0.00000002 │
│  Combined power               0.94042720 │
│                                     │
│     [ Assistant ]   [ Compute ]     │
└─────────────────────────────────────┘
```

The user can enter alpha, tails, $df_{Denominator}$, and the non-centrality parameter (NCP). The program displays the t-value required for significance and power.

For a one-sample test, NCP $= d\sqrt{N}$, and $df_{Denominator} = N - 1$.

For a two-sample test, NCP $= \dfrac{d\sqrt{n'}}{\sqrt{2}}$, where n' is the harmonic mean of N1 and N2, and $df_{Denominator} = Ntotal - 2$.

The program also offers assistant panels. When the assistant is active, values for degrees of freedom (df) and NCP must be entered through the assistant. To enter these values directly, close the assistant.

Figure 20.2 Non-central t (one group) with assistant

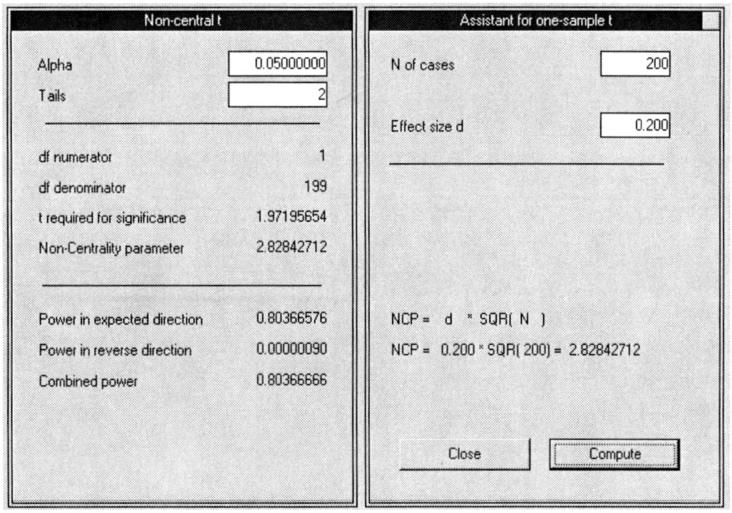

Non-central t	
Alpha	0.05000000
Tails	2
df numerator	1
df denominator	199
t required for significance	1.97195654
Non-Centrality parameter	2.82842712
Power in expected direction	0.80366576
Power in reverse direction	0.00000090
Combined power	0.80366666

Assistant for one-sample t	
N of cases	200
Effect size d	0.200

NCP = d * SQR(N)
NCP = 0.200 * SQR(200) = 2.82842712

The assistant panel for one-sample t-tests is shown in Figure 20.2.

▶ Enter the number of cases.

▶ Enter the effect size, d.

▶ Enter alpha and tails.

The program displays:
- The computed non-centrality parameter
- The required t
- Power

Figure 20.3 Non-central t (two-group) with assistant

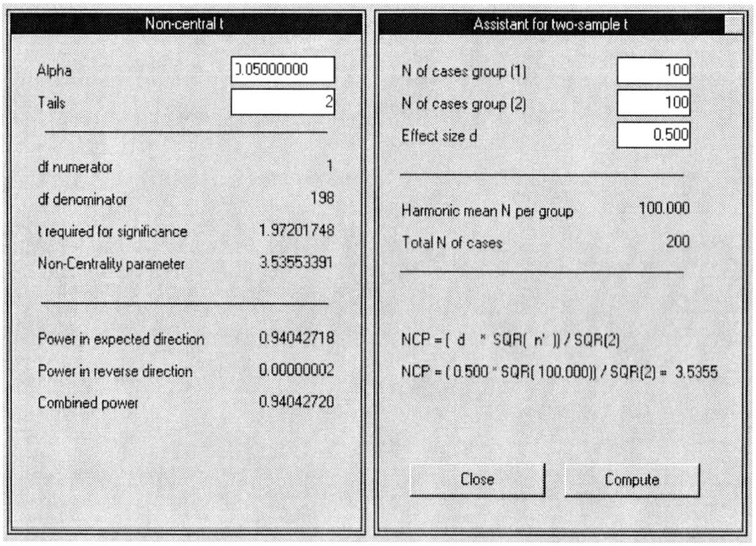

Non-central t		Assistant for two-sample t	
Alpha	0.05000000	N of cases group (1)	100
Tails	2	N of cases group (2)	100
		Effect size d	0.500
df numerator	1		
df denominator	198	Harmonic mean N per group	100.000
t required for significance	1.97201748	Total N of cases	200
Non-Centrality parameter	3.53553391		
Power in expected direction	0.94042718	NCP = (d * SQR(n')) / SQR(2)	
Power in reverse direction	0.00000002	NCP = (0.500 * SQR(100.000)) / SQR(2) = 3.5355	
Combined power	0.94042720		

The assistant panel for two-sample t-tests is shown in Figure 20.3.

▶ Enter the number of cases.

▶ Enter the effect size, d.

▶ Enter alpha and tails.

The program displays:
- The computed non-centrality parameter
- The required t
- Power

General Case (Non-Central F)

Figure 20.4   Non-central F (ANOVA)

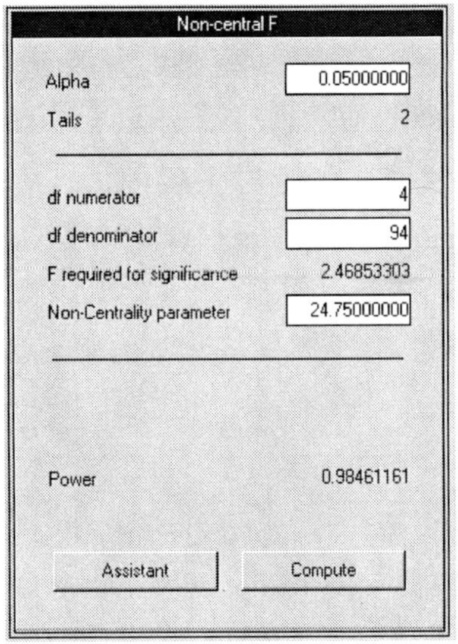

Alpha	0.05000000
Tails	2
df numerator	4
df denominator	94
F required for significance	2.46853303
Non-Centrality parameter	24.75000000
Power	0.98461161

The general case for power based on non-central F is shown in Figure 20.4.

The user can enter alpha, tails, $df_{Numerator}$, $df_{Denominator}$, and the non-centrality parameter (NCP). The program displays the F-value required for significance and power.

NCP is computed as $f^2 * (df_{Numerator} + df_{Error} + 1)$. In the case of a oneway ANOVA or a multiple regression with only one set of variables, this is equivalent to $f^2 * N_{total}$. In other cases, however, the two formulas are not identical because the term in parentheses does not include the degrees of freedom (df) associated with other factors or covariates (in a factorial ANOVA) or other sets of variables (in a multiple regression).

The program also offers assistant panels for ANOVA and multiple regression. In these panels, the user can enter data that are used in intermediate computations and transferred to the first panel. When the assistant is active, values for df and NCP must be entered through the assistant. To enter these values directly, close the assistant.

Figure 20.5 Non-central F (ANOVA) and assistant

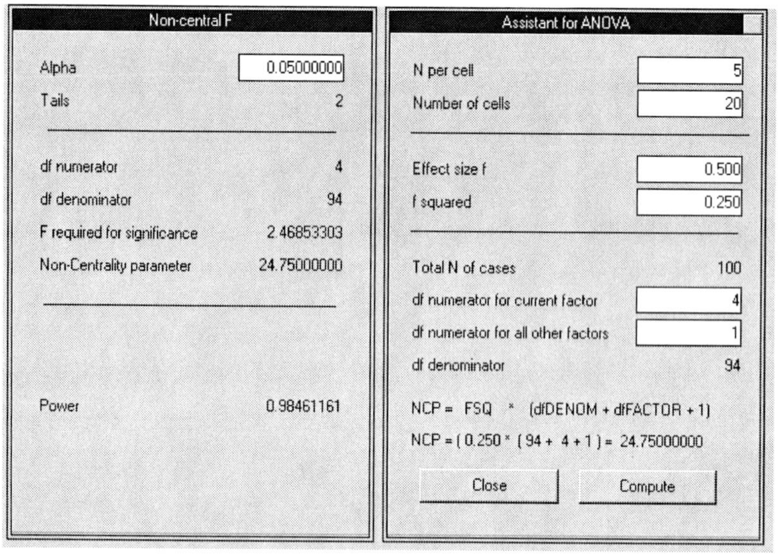

Non-central F	
Alpha	0.05000000
Tails	2
df numerator	4
df denominator	94
F required for significance	2.46853303
Non-Centrality parameter	24.75000000
Power	0.98461161

Assistant for ANOVA	
N per cell	5
Number of cells	20
Effect size f	0.500
f squared	0.250
Total N of cases	100
df numerator for current factor	4
df numerator for all other factors	1
df denominator	94

NCP = FSQ * (dfDENOM + dfFACTOR + 1)
NCP = (0.250 * (94 + 4 + 1) = 24.75000000

The assistant panel for ANOVA is shown in Figure 20.5.

▶ Enter alpha.

▶ Enter the number of cases per cell and the number of cells.

▶ Enter the effect size, f, or f^2.

▶ Enter the $df_{Numerator}$ for the factor of interest.

▶ Enter the $df_{Numerator}$ for all other factors and covariates.

▶ The program computes the $df_{Denominator}$ as $Ntotal - df_{Factor} - df_{Other}$.

▶ Click *Compute*.

The program displays:
- The F-value required for significance
- The non-centrality parameter
- Power

General Case

Figure 20.6 Non-central F (multiple regression) and assistant

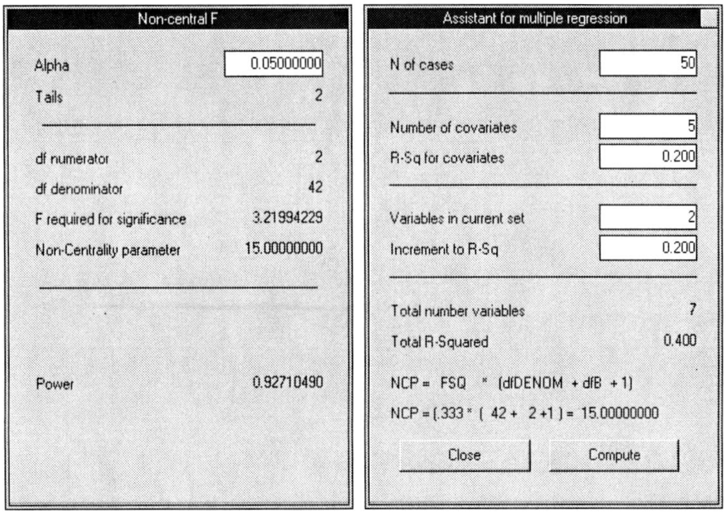

Non-central F	
Alpha	0.05000000
Tails	2
df numerator	2
df denominator	42
F required for significance	3.21994229
Non-Centrality parameter	15.00000000
Power	0.92710490

Assistant for multiple regression	
N of cases	50
Number of covariates	5
R-Sq for covariates	0.200
Variables in current set	2
Increment to R-Sq	0.200
Total number variables	7
Total R-Squared	0.400
NCP = FSQ * (dfDENOM + dfB +1)	
NCP = (.333 * (42 + 2 +1) = 15.00000000	

The assistant panel for multiple regression is shown in Figure 20.6.

▶ Enter alpha.

▶ Enter the number of cases.

▶ Enter the number of covariates and the R^2 for these covariates.

▶ Enter the number of variables in the current set and the increment to R^2 for the current set.

The program displays:
- The F-value required for significance
- f^2
- The non-centrality parameter
- Power

General Case (Non-Central Chi-Square)

Figure 20.7 Non-central chi-square

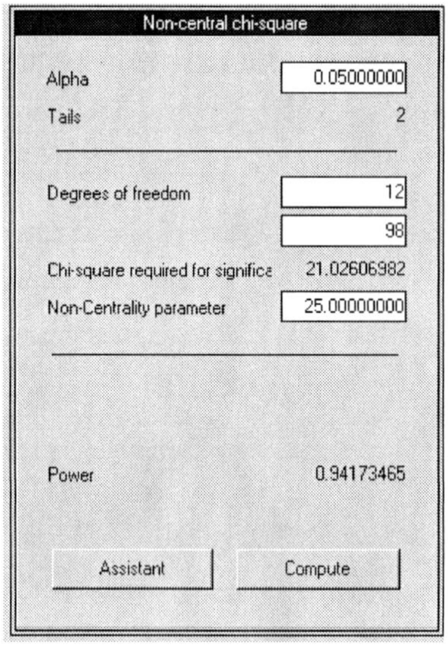

Non-central chi-square	
Alpha	0.05000000
Tails	2
Degrees of freedom	12
	98
Chi-square required for significance	21.02606982
Non-Centrality parameter	25.00000000
Power	0.94173465

The general case for power based on non-central chi-square is shown in Figure 20.7.

The user can enter alpha, tails, degrees of freedom (df), and the non-centrality parameter (NCP). The program displays the chi-square value required for significance and power.

NCP is computed as $w^2 * N$.

The program allows the user to enter the NCP directly. The program also allows the user to invoke an assistant that accepts data in the form of a $K \times C$ crosstabulation, which it then uses to compute the NCP. The assistant shows the steps in the computation. When the assistant is active, values for df and NCP must be entered through the assistant. To enter these values directly, close the assistant.

Figure 20.8 Non-central chi-square and assistant

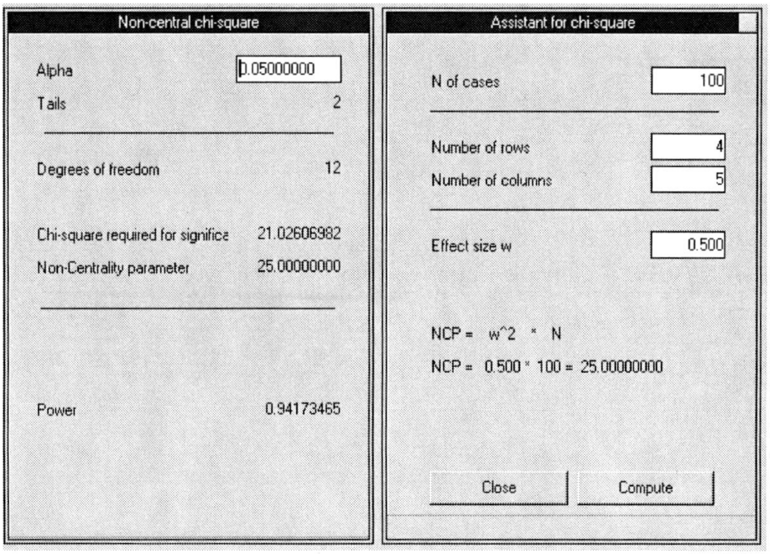

The assistant panel for non-central chi-square is shown in Figure 20.8.

▶ Enter alpha.

▶ Enter the number of cases.

▶ Enter the number of rows and columns in the K × C crosstabulation.

▶ Enter the effect size, w.

The program displays:
- The computed w^2
- The non-centrality parameter
- The chi-square value required for significance
- Power

Printing the General Case Panel

To print the panel, click the *Print* icon, or choose *Print* from the File menu. Only the panel on the left side is printed.

Copying Data to the Clipboard

To copy data to the clipboard, click the *Copy to clipboard* icon, or choose *Clipboard* from the File menu. Only the panel on the left side is copied to the clipboard.

Appendix A
Installation

Requirements

This program will run under Windows 95.

Installing

On the Windows 95 Start menu, use the *Run* command to run setup. Or, click *Settings* to access the Control Panel, and then double-click *Add/Remove Programs*.

Uninstalling

The program folder will include an uninstall program. This program will prompt you before deleting any files that may be shared by other programs. Do not delete these files unless you are certain that they are not needed.

Troubleshooting

Before installing the program, close any other active programs, including the Microsoft Office Shortcut Bar.

Note

The program installs and registers the following files into the Windows *system* directory (or the Windows *sys32* directory) and overwrites any earlier versions of the same files:

mfcans32.dll, oc30.dll, oleaut32.dll, olepro32.dll, csmete32.ocx, csspin32.ocx, cstext32.ocx, ccmsghk.dll, imghook.dll, tmghook.dll, ss32x25.ocx, threed32.ocx, vsocx32.ocx, vsview2.ocx, smthlp32.ocx, vsflex2.ocx, richtx32.ocx, gswdll32.dll, gswag32.dll, graphs32.ocx, vb40032.dll, ven2232.olb, msvcrt40.dll, msvcrt.dll, mfc40.dll, olepro32.dll, comdlg32.ocx, graph32.ocx, gsw32.exe, gswdll32.dll

The program installs and registers the following files into the Windows *system* directory (or the Windows *sys32* directory) but does *not* overwrite earlier versions of the same files:

comctl32.ocx, comctl32.dll

The following files are written to the Windows *help* directory:

power.hlp, power.cnt

Additional files are written to the application directory.

All activities of the installation program are written to a log file in the program directory.

Appendix B
Troubleshooting

Following are explanations for problems that may occur.

Table, graph, or report does not update

The program will update the table, graph, or report if power has changed since the previous update. In some cases, you will want to force an update (such as when the means are changed in a way that does not affect power). In this case, choose *Force recompute* from the Tools menu.

Power will not rise above some point

When power is being computed for a two-sample study (such as a two-sample t-test) and cases are allocated unevenly to the two groups, the effective number of cases is driven primarily by the smaller of the two sample sizes. There is a maximum effective number of cases, defined as a function of the lesser number of cases, that will not be exceeded no matter how much the size of the other group is increased.

Required power reverts to 90%

Changes to required power remain in effect only as long as the user remains within a given module. To set another value as the default, choose *Preferences* from the Options menu.

Graph lines appear jagged

Set the table to display more decimal places (3 is a good starting point).

Use a smaller increment for the number of cases (the program can set the range and increment automatically).

Graph menu shows power only (not precision)

Set the style to power *and* precision (choose *Style* from the Modify menu).

Graph menu doesn't include power as function of effect size or alpha

Set the style to power *only* (choose *Style* from the Modify menu).

Graph menu doesn't include standard error of index

Some indices, such as the odds ratio, have confidence intervals that are markedly asymmetric. The program will graph the confidence intervals for these (showing the asymmetry) but will not graph the standard error because this value is not generally meaningful.

An increase in sample size results in a reduction in power

This is possible for exact tests of proportions because the increase in sample size may yield a more conservative value for alpha.

Appendix C
Computational Algorithms for Power

Computation of Power

The process of computing power varies from one statistical procedure to the next but always follows the same logic.

First, we determine what value of the test statistic will be required to yield a significant effect. For example, if we are planning a t-test with N = 20 per group and alpha (2-tailed) set at 0.05, then a specific t-value will be required to yield a significant effect. This value is a function of the level of significance (alpha), the number of tails, and (for many tests) the sample size.

Second, we determine what proportion of studies is expected to yield a test statistic that meets this criterion. This value, which is the study's power, is a function of the effect size under the alternate hypothesis, and the sample size.

The specific computations required for these two steps are outlined here for each of the program's modules. References are to IMSL subroutines.

The operation of these algorithms can be observed directly by selecting the General panel from the procedures available (choose *New analysis* from the File menu). This allows the user to enter the non-centrality parameter (NCP) and degrees of freedom (df) directly or through "assistant" screens. In this module, the program displays the intermediate results of calculations so that the steps outlined here, and the computed value at each step, can be observed.

T-Test (One-Sample and Paired) with Estimated Variance

Required T and Power

The t-value required for significance (ReqT) is given by the central t-distribution for significance level alpha and/or tails and df = N − 1. The value is computed using the DTIN algorithm.

The effect size (d) is computed as

d = ABS(Mean Difference) / SD

The non-centrality parameter (NCP) is computed as NCP = Abs(d * Sqrt(N_{Cases})).
Power is given by the non-central t for NCP, ReqT, and df. The value is computed using the DTDF algorithm.

One-Sample Test and Paired T-Test

The same formula is applied for the one-sample t-test and the paired t-test. For the paired t-test, the program allows the user to enter the standard deviation for each time point and computes the standard deviation of the difference using the formula

SD_{Diff} = Sqrt([SD(1)]² + [SD(2)]² − (2 * Corr * SD(1) * SD(2)))

One-Tailed versus Two-Tailed Tests

For a one-tailed test, ReqT is found for alpha, and power is computed only for an effect in one direction (using the absolute value of d).

For a two-tailed test, ReqT is found for alpha/2. Power is computed for an effect in one direction using the absolute value of NCP, it is computed for an effect in the reverse direction using −1 * NCP, and the two values are summed.

T-Test (Two-Sample) with Estimated Variance

Required T and Power

The t-value required for significance (ReqT) is given by the central t-distribution for significance level alpha and/or tails and df = N1 + N2 − 2. The value is computed using the DTIN algorithm.

The effect size (d) is computed as d = ABS(Mean Difference) / SD_{Pooled}, where

SD_{Pooled} = Sqrt((((N1 − 1) * SD1 * SD1 + (N2 − 1) * SD2 * SD2) / (N1 + N2 − 2)))

The harmonic mean of the number of cases is given by

HarmonicN = (2 * N1 * N2) / (N1 + N2)

The non-centrality parameter is computed as

NCP = Abs((d * Sqrt(HarmonicN)) / Sqrt(2))

Power is given by the non-central t for NCP, ReqT, and df. The value is computed using the DTDF algorithm.

One-Tailed versus Two-Tailed Tests

For a one-tailed test, ReqT is found for alpha, and power is computed only for an effect in one direction (using the absolute value of d).

For a two-tailed test, ReqT is found for alpha/2. Power is computed for an effect in one direction using the absolute value of NCP, it is computed for an effect in the reverse direction using –1 * NCP, and the two values are summed.

Z-Test (One-Sample and Paired) with Known Variance

Required Z and Power

The z-value required for significance (Z_{req}) is given by the normal distribution for significance level alpha and/or tails. The value is computed using the DNORIN algorithm.

The effect size (d) is computed as

d = Abs(Mean Difference) / SD

The non-centrality parameter (NCP) is computed as

NCP = Abs(d * Sqrt(NCases))

Power is given by the normal distribution for NCP and Z_{req}. The value is computed using the DNORDF algorithm.

One-Sample Test and Paired T-Test

The same formula is applied for the one-sample t-test and the paired t-test. For the paired t-test, the program allows the user to enter the standard deviation for each time point and computes the standard deviation of the difference using the formula

SD_{Diff} = Sqrt([SD(1)]^2 + [SD(2)]^2 – (2 * Corr * SD(1) * SD(2)))

One-Tailed versus Two-Tailed Tests

For a one-tailed test, Z_{req} is found for alpha, and power is computed only for an effect in one direction (using the absolute value of d).

For a two-tailed test, Z_{req} is found for alpha/2. Power is computed for an effect in one direction using the absolute value of NCP, it is computed for an effect in the reverse direction using –1 * NCP, and the two values are summed.

In this description, we have used the term "non-central" to maintain the parallel with the t-test, although the normal distribution is symmetrical.

Z-Test (Two-Sample) with Known Variance

Required Z and Power

The z-value required for significance (Z_{req}) is given by the normal distribution for significance level alpha and/or tails. The value is computed using the DNORIN algorithm.

The effect size (d) is computed as

d = Abs(Mean Difference) / SDPooled

where

SDPooled = Sqrt((((N1 – 1) * SD1 * SD1 + (N2 – 1) * SD2 * SD2) / (N1 + N2 – 2)))

The harmonic mean of the number of cases is given by

HarmonicN = (2 * N1 * N2) / (N1 + N2)

The non-centrality parameter is computed as

NCP = Abs((d * Sqrt(HarmonicN)) / Sqrt(2))

Power is given by the normal distribution for NCP and Z_{req}. The value is computed using the DNORDF algorithm.

One-Tailed versus Two-Tailed Tests

For a one-tailed test, Z_{req} is found for alpha, and power is computed only for an effect in one direction (using the absolute value of NCP).

For a two-tailed test, Z_{req} is found for alpha/2. Power is computed for an effect in one direction using the absolute value of NCP, it is computed for an effect in the reverse direction using $-1 * NCP$, and the two values are summed.

Single Proportion versus a Constant

Single Proportion versus Constant: Arcsin Method

The arcsin transformation of a proportion is defined as

Arcsin = 2 * Atn(x / Sqrt(–x * x + 1))

where

x	=	Sqrt(p)
H1	=	Arcsin transformation of P1
H2	=	Arcsin transformation of P2
HDIFF	=	H1 – H2
Z_{req}	=	Z-value required for significance
N	=	N of cases
Z_{power}	=	Abs(HDIFF * Sqrt(2)) * Sqrt(N / 2) – Z_{req}
Power	=	Computed for Z_{power} based on normal distribution

One-Tailed versus Two-Tailed Tests

For a one-tailed test, Z_{req} is found for alpha, and power is computed only for an effect in one direction (using the absolute value of HDIFF).

For a two-tailed test, Z_{req} is found for alpha/2. Power is computed for an effect in one direction using the absolute value of HDIFF, it is computed for an effect in the reverse direction using $-1 * Abs(HDIFF)$, and the two values are summed.

Single Proportion versus a Constant (Exact Test)

As noted above, in all procedures the program first finds the critical value required for significance under the null and then finds the proportion of studies (under the alternate) that will meet this criterion. In the exact test for proportions, both steps in this process require a consideration of all possible discrete outcomes.

The program uses the binomial distribution to find the number of successes required to yield significance at the nominal value of alpha. After establishing the number of successes required to yield a significant effect, the program again uses the binomial distribution to find the proportion of cases expected to meet this criterion, given the "true" success rate under the alternate hypothesis.

Power for an effect in the other direction is computed by the analogous process—we find the first number of successes that will allow us to reject the hypothesis, and then determine what proportion of cases will meet this criterion.

One-Tailed and Two-Tailed Tests

For a one-tailed test, the criterion alpha used in the first step is the alpha specified by the user. The program computes power for an effect in either direction and reports only the higher of these two values. The program also reports the actual alpha.

For a two-tailed test, the criterion alpha used in the first step is alpha/2. The program computes power for an effect in either direction and sums these values. Similarly, actual alpha is the sum of actual alpha in either direction.

Two Independent Proportions

Two Proportions: Arcsin Method

The arcsin transformation of a proportion is defined as

Arcsin = 2 * Atn(x / Sqrt(–x * x + 1))

where

x	=	Sqrt(p)
H1	=	Arcsin transformation of P1
H2	=	Arcsin transformation of P2
HDIFF	=	H1 – H2
Z_{req}	=	Z-value required for significance
N'	=	(2 * (N1 * N2)) / (N1 + N2)
Z_{power}	=	Abs(HDIFF) * Sqrt(N' / 2) – Z_{req}
Power	=	Computed for Z_{power} based on anormal distribution

One-Tailed versus Two-Tailed Tests

For a one-tailed test, Z_{req} is found for alpha, and power is computed only for an effect in one direction (using the absolute value of HDIFF).

For a two-tailed test, Z_{req} is found for alpha/2. Power is computed for an effect in one direction using the absolute value of HDIFF, it is computed for an effect in the reverse direction using $-1 * Abs(HDIFF)$, and the two values are summed.

Two Proportions: Normal Approximations

Assume that P1 is the higher of the two proportions and P2 is the lesser of the two.

N'   = (2 * N1 * N2) / (N1 + N2)
Z_{req} = Z required for significance, given alpha
$Mean_P$ = (P1 + P2) / 2
Q1   = 1 – P1
Q2   = 1 – P2
$Mean_Q$ = 1 – $Mean_P$
d'   = 0.5 + Z_{req} * Sqrt(0.5 * N' * $Mean_P$ * $Mean_Q$)
Zpower = (d' – 0.5 * N' * (P1 – P2) – c) / Sqrt(0.25 * N' * (P1 * Q1 + P2 * Q2))

Power is found from the normal distribution for Zpower.
　The program incorporates three versions of this approximation, which differ only in the definition of the correction factor, c.
　For the method identified simply as the normal approximation with unweighted $Mean_P$, c = 0.5.
　For the Kramer-Greenhouse method, c = – 0.5.
　For the Casagrande and Pike method, c = 0.

Computation of $Mean_P$ under the Null

Power is always computed based on the variance expected under the null, and in a test of two proportions the variance under the null is a function of P1 and P2 (that is, the proportion of positive cases in either group).
　The methods identified as the normal approximation, Kramer-Greenhouse, and Casagrande and Pike all use (P1+P2)/2 to compute the common population proportion (and variance) under the null.
　A variant on this formula uses the weighted mean of P1 and P2 to compute the proportion under the null. This option is not generally appropriate, but the method is discussed in the literature and is offered here for special cases.

$Mean_P$ = Abs((Cell(1, 3) * P1 + Cell(2, 3) * P2) / (Cell(1, 3) + Cell(2, 3)))

where the cell is identified by Cell(row,column). If N1 = N2, this will yield identical results to the method identified as the normal approximation with unweighted $Mean_P$.

One-Tailed versus Two-Tailed Tests

For a one-tailed test, Z_{req} is found for alpha, and power is computed only for an effect in one direction by assigning the higher proportion to P1 in the algorithm.

For a two-tailed test, Z_{req} is found for alpha/2. Power is computed for an effect in one direction using P1 > P2, it is computed for an effect in the reverse direction by assigning the lower proportion to P1 in the algorithm, and the two values are summed.

Two Proportions—Chi-Square Test

The user provides the sample size in each group and the proportion of "positive" cases in each group. The program uses this information to create the corresponding 2×2 table, corresponding to the alternate hypothesis.

The chi-square value required for significance is given by the DCHIIN algorithm for alpha and df = 1.

The non-centrality parameter (NCP) is defined as $w^2 * N_{total}$, where w is defined as in the K × C analysis. The NCP defined in this way is equivalent to the chi-square value computed for the 2×2 table, and the program works with this value, which is given by the CTTWO algorithm.

The program allows the user to select chi-square with the Yates correction. When this option is selected, the NCP is adjusted and is given the Yates-corrected chi-square value returned by the CTTWO algorithm.

Power is then given by the non-central chi-square distribution for df = 1, required chi-sq, and NCP. This value is computed by the DSNDF algorithm.

One-Tailed versus Two-Tailed Tests

The chi-square distribution is one-tailed (corresponding to a nondirectional test). When this computational option is selected, the program will compute power for a nondirectional test only.

Two Proportions: Fisher's Exact Test

The user provides the number of cases in each row and the proportion of "positive" cases in each row under the alternate hypothesis.

The program iterates through every possible outcome for row 1. For example, if 10 cases will be assigned to row 1, it is possible to obtain 0, 1, 2, ..., 10 successes in this row. A likelihood value is assigned to each row using the binomial distribution, taking into account the proportion of successes under the alternate.

Since the two rows are independent of each other, the probability of any given joint outcome is computed as the product of the likelihood for a given outcome in row 1 by a given outcome in row 2. For example, if the likelihood of 5 failures/5 successes in row 1 is 10% and the likelihood of 2 failures/8 successes in row 2 is 2%, the likelihood of drawing a sample with 5/5 in row 1 and 2/8 in row 2 is 0.10 * 0.02 = 0.002.

The program iterates through every possible combination of outcomes. For each combination, the program computes Fisher's exact test and records the p-value. If the p-value is equal to or less than alpha, the likelihood of this outcome is added to a cumulative total; at the conclusion of the iterative process, this cumulative total gives the proportion of outcomes that yield significance—that is, power.

One-Tailed versus Two-Tailed Tests

The algorithm used to compute significance for any 2×2 table computes both a one-tailed and a two-tailed p-value.

When power is computed for a one-tailed test, the one-tailed p-value is used but is counted only if the effect is in the expected direction. When power is computed for a two-tailed test, the two-tailed p-value is used.

To speed up computations, the program does not compute the Fisher exact test for 2×2 tables that are very unlikely to occur (defined as likelihood < 0.0000001). Even if this combination yielded a significant result, the increment to power would be 0.0000001 or less.

McNemar Test of Paired Proportions

The McNemar test is a special case of one proportion against a constant. Cases are assigned to the four cells:

	Negative	Positive
Negative	AA	AB
Positive	BA	BB

For example, we might specify that cases will fall into the four cells in the following proportions:

	Negative	Positive
Negative	15%	30%
Positive	20%	35%

Information about the treatment effect is provided by cells AB and BA, while cells AA and BB provide no information.

The effect size is based solely on cells AB+BA. Concretely, we work with 30% versus 20%, or 50% of the sample in this example. The proportion of cases falling into cell AB is computed based on this subsample as AB/(AB+BA). In this example, 30/50 = 60% fall into cell AB. Under the null hypothesis, this value will always be 50%.

The adjusted proportion in cell AB (60%) is tested against the constant (50%) using either of the formulas described for one proportion against a constant (the arcsin approximation or the exact test). In either case, the number of cases is given by the N in these two cells.

Sign Test

The sign test is used to test the hypothesis that the proportion of cases falling into two groups is equal (that is, that the proportion in either group is 50%).

This test is a special case of a single sample proportion against a constant. The proportion of cases falling into cell A is tested against a constant of 50% using either of the formulas described for one proportion against a constant (the arcsin approximation or the exact test). In either case, the number of cases is given by the full N.

K x C Crosstabulation

Computing Effect Size

The table provided by the user (which requires that the cells in a row sum to 1.0, and the proportion of cases falling into each row is given separately) is used to create a table in which the value in each cell is that cell's proportion of the total sample, so that the cells in the table (rather than the row) sum to 1.0, the row marginals sum to 1.0, and the column marginals sum to 1.0.

Under the null hypothesis, the expected value in each cell is defined as the product of that cell's marginals. For each cell, the discrepancy between the null and the alternate is computed as ((PercentAlternate−PercentNull)^2/PercentNull). The sum of these squared discrepancies yields w^2, and the square root of this value yields the effect size, w.

Computing Power

The degrees of freedom is computed as (Rows − 1) * (Columns − 1).

The chi-square value required for significance is given by the central chi-square distribution for alpha and df. This value is computed by the DCHIIN algorithm.

The non-centrality parameter (NCP) is computed as $w^2 * N_{Total}$.

Power is given by the non-central chi-square distribution for required chi-sq, NCP, and df. This value is computed by the DSNDF algorithm.

One-Tailed versus Two-Tailed Tests

The K × C test is nondirectional.

Correlations—One Sample

The test of a single correlation against a null hypothesis of 0 is computed by the exact method, which will yield the same value as the multiple regression procedure. Tests of a single correlation versus a non-zero constant and tests that two correlations are different from each other are carried out using the Fisher-Z transformation.

Pearson Correlation—One Sample versus Zero

The t-value required for significance (ReqT) is given by the central t-distribution for significance level alpha and/or tails and df = $N_{Total} - 2$. The value is computed using the DTIN algorithm.

The user provides the correlation coefficient, r, which is squared to yield r^2.

The non-centrality parameter (NCP) is computed as

NCP = t * Sqrt(NCases)

where

t = Sqrt(r^2 / (1 – r^2))

Power is given by the non-central t distribution for NCP, ReqT, and df. The value is computed using the DTDF algorithm.

One-Tailed versus Two-Tailed Tests

For a one-tailed test, ReqT is found for alpha, and power is computed only for an effect in one direction (using the absolute value of NCP).

For a two-tailed test, ReqT is found for alpha/2. Power is computed for an effect in one direction using the absolute value of NCP, it is computed for an effect in the reverse direction using –1 * NCP, and the two values are summed.

One Sample versus Constant Other than Zero

In this description, Zr is used to denote the Fisher-Z transformation of r.

Zr_1 is the Fisher-Z transformation of the correlation under the alternate, and Zr_2 is the Fisher-Z transformation of the constant against which r_1 will be tested.

The effect size, Q, is computed as $Q = Abs(Zr_1 - Zr_2)$.

N is the number of cases.

Z_{req} (the z-value required for significance) is Z_p, which is the z-value corresponding to power (not to be confused with Fisher's Z). Z_p is computed as

$$Z_p = Q * Sqrt((N - 3) / 1) - Z_{req}$$

Power is then computed as the area under the curve to the left of Z_p using the normal distribution.

One-Tailed versus Two-Tailed Tests

For a one-tailed test, Z_{req} is found for alpha, and power is computed only for an effect in one direction (using Q).

For a two-tailed test, Z_{req} is found for alpha/2. Power is computed for an effect in one direction using Q, it is computed for an effect in the reverse direction using $-1 * Q$, and the two values are summed.

Correlations—Two Sample

In this description, Zr is used to denote the Fisher-Z transformation of r. Zr_1 is the Fisher-Z transformation of the first correlation, and Zr_2 is the Fisher-Z transformation of the second correlation.

The effect size, Q, is computed as $Q = Abs(Zr_1 - Zr_2)$.

$$N' = ((2 * (N1 - 3) * (N2 - 3)) / (N1 + N2 - 6)) + 3$$

Z_{req} (the z-value required for significance) is Z_p, which is the z-value corresponding to power (not to be confused with Fisher's Z). Z_p is computed as

$$Z_p = Q * Sqrt((N' - 3) / 2) - Z_{req}$$

Power is then computed as the area under the curve to the left of Z_p using the normal distribution.

One-Tailed versus Two-Tailed Tests

For a one-tailed test, Z_{req} is found for alpha, and power is computed only for an effect in one direction (using Q).

For a two-tailed test, Z_{req} is found for alpha/2. Power is computed for an effect in one direction using Q, it is computed for an effect in the reverse direction using $-1 * Q$, and the two values are summed.

Analysis of Variance

The program will compute power for a oneway analysis of variance, or a balanced factorial ANOVA, for a fixed effects model.

Computing the Effect Size (f)

The effect size (f) is computed in one of four ways.

The program allows the user to enter f directly, in which case the value given for SD_{Within} has no impact on f.

The user can enter the $SD_{Between}$ levels for a factor, in which case the program computes

$f = SD_{Between} / SD_{Within}$

The user can enter the mean for each level in a factor, in which case the program computes the $SD_{Between}$ levels for a factor, and then computes

$f = SD_{Between} / SD_{Within}$

Finally, the user can enter the range of means (that is, the single lowest and highest means) and the pattern of dispersion for the remaining means. In this case, we compute the effect size (d) for the two extreme groups, using the same method as for a t-test.

$d = (Mean_{HI} - Mean_{LO}) / SD_{Within}$

Then, this value is adjusted to take account of the pattern of remaining means.

If the other means are clustered at the center,

$f = d * Sqrt(1 / (2 * k))$

If the means are spread uniformly over the range,

f = (d / 2) * Sqrt((k + 1) / (3 * (k – 1)))

If the means are at the extremes, we need to distinguish between two cases:
- If the number of groups is even, f = 0.5 * d.
- If the number of groups is odd, f = d * (Sqrt((k * k) – 1)) / (2 * k).

Computing the F-Value Required for Significance

The F-value required for significance (ReqF) is given by the central F distribution for significance level alpha and DF_1, DF_2 where DF_1 is degrees of freedom for the current factor (or interaction) and

$$DF_2 = N_{total} - DF_{Factor} - DF_{OtherFactors} - DF_{Interactions} - DF_{Covariates}$$

This value is obtained by the DFIN algorithm.

For each factor or interaction, the value of f computed above is squared to yield f^2, which is used in computing the non-centrality parameter (NCP):

$$NCP = f^2 * (DF_1 + DF_2 + 1)$$

Equivalently,

$$NCP = f^2 * (N_{total} - DF_{other})$$

where DF_{other} is the df associated with factors (or interactions) other than the current one, or with covariates.

Power is given by the non-central F distribution for NCP, ReqF, DF_1, and DF_2. The value is computed using the DFFNCD algorithm.

Note. The computation of NCP is based on the effect size and the DF's that are available for the current test—that is, DF_{error} and $DF_{Numerator}$. DF's that have been assigned to other factors or interactions are excluded from this computation. It is possible to compute NCP as $f^2 * N_{total}$. But this would seem to be appropriate only when other factors and interactions are assumed to explain no variance and are therefore pooled with the error term. The formula used by this program is appropriate for the vast majority of cases and will be slightly conservative otherwise.

The algorithm adopted for computing power for ANOVA (and ANCOVA) is consistent with the algorithm used for multiple regression, where computation of NCP is similarly based on DF_{error} and DF for the current factor, with DF for other factors excluded.

One-Tailed versus Two-Tailed Tests

Analysis of variance is nondirectional.

Analysis of Covariance

Power computation for analysis of covariance (ANCOVA) proceeds exactly as for ANOVA except that the effect size and df are modified.

The effect size is adjusted as follows:

$$f_{adjusted} = f / Sqrt(1 - R^2_{cov})$$

Additionally, the degrees of freedom attributed to the covariate are removed from DF_{error}, which affects the computation of ReqF and the NCP. Recall that the NCP is computed as $f^2 * (DF_1 + DF_2 + 1)$. The inclusion of a covariate will increase the first term but decrease the second term (since DF_2 will be lowered). Unless the sample size is quite small, the impact on f^2 will tend to overwhelm the impact on DF_2.

Multiple Regression

Definition of Terms

N_{total}	Total number of cases
Set A	Set of covariates
K_A	Number of variables in set A
I^2_A	Increment to R^2 for set A
Set B	Set for which power will be computed
K_B	Number of variables in set B
I^2_B	Increment to R^2 for set A
Set C	Set of additional covariates, entered subsequent to set B
K_C	Number of variables in set C
I^2_C	Increment to R^2 for set C

Note that the number of variables in set A and/or C may be set to 0, so this construction can be used to define any possible multiple regression analysis.

The program displays the multiple regression as a series of lines, each representing a set of variables. For each line it gives power for the increment due to that line and also for the cumulative impact of all variables through the current line.

For analysis of line 1, set A is defined as nil, set B is defined as line 1, and set C is defined as lines 2–10.

For analysis of line 2, set A is defined as line 1, set B is defined as line 2, and set C is defined as lines 3–10.

For analysis of line 3, set A is defined as lines 1–2, set B is defined as line 3, and set C is defined as lines 4–10. The same approach is used for lines 4–10.

The program also allows the user to designate a set that consists of more than one line, say lines 4–6. In this case, set A is defined as lines 1–3, set B is lines 4–6, and set C is lines 7–10.

Model 1 versus Model 2 Error

Under model 2 error (the default), all sets (including set C) are assumed to be in the model at every point. Thus, for example, when power is computed for line 2 in a five-line model, the R^2_{Total} is based on sets A, B, and C (in effect, the cumulative R^2 through line 5), and the Df_{error} assumes that all five lines are in the model. Under model 1 error, set C is always nil. In other words, the analysis of each line assumes that all subsequent lines are empty. (Model 1/model 2 error should not be confused with 1-tail/2-tail tests or with type 1/type 2 error).

Computation of Power

$$R^2_{Total} = I^2_A + I^2_B + I^2_C$$
$$F^2 = I^2_B / (1 - R^2_{Total})$$

The F-value required for significance (ReqF) is given by the central F distribution for significance level alpha and DF_1, DF_2 where

$$DF_1 = K_B$$
$$DF_2 = N_{Cases} - K_A - K_B - K_C - 1$$

This value is obtained by the DFIN algorithm.

The non-centrality parameter is computed as

$$NCP = f^2 * (DF_1 + DF_2 + 1)$$

Equivalently,

$$\text{NCP} = f^2 \ast (N_{Total} - K_A - K_C)$$

Power is given by the non-central F distribution for NCP, ReqF, DF_1, and DF_2. The value is computed using the DFFNCD algorithm.

One-Tailed versus Two-Tailed Tests

The multiple regression test is nondirectional.

Appendix D
Computational Algorithms for Precision

T-Test for Means (One-Sample) with Estimated Variance

The width of the confidence interval computed for a *completed* study depends on the confidence level, the standard deviation, and the sample size. The width of the confidence interval for a *planned* study depends on these values and on the sampling distribution of the standard deviation as well.

The t-value corresponding to the confidence level (tCI) is given by the central t-distribution for $(1 - CI_{Level})$ / Tails for $df = N - 1$. The value is computed using the DTIN algorithm.

The standard error of the mean is given by

SE = SD1 / Sqrt(N)

The lower and upper limits of the confidence interval are given by

LowerLimit = Mean1 − t_{CI} * SE * 50% Tolerance Factor
UpperLimit = Mean1 + t_{CI} * SE * 50% Tolerance Factor

where the 50% Tolerance Factor = Sqrt(CHI_{Req} / df) for $df = N - 1$. This value is given by the function DCHIIN for 1 − Tolerance level, and df, and takes account of the sampling distribution of the standard deviation.

The value computed in this way is the median value. Half the studies will yield a confidence interval wider than the expected value and roughly half will yield a narrower interval.

One-Tailed versus Two-Tailed Tests

Assuming the 95% confidence level, the t-value computed for a one-tailed test would be based on 1 – 0.95 (that is, 0.05). The t-value corresponding to a two-tailed test would be based on (1 – 0.95)/2 (that is, 0.025).

The one-tailed "interval" yields a lower (or upper) limit closer to the observed value than does the two-tailed interval, but only one of the limits is meaningful. The one-tailed "interval" extends from minus infinity to the upper limit reported, or from the lower limit reported to positive infinity.

Tolerance intervals

The program is able to compute also the tolerance interval for a given confidence interval; that is, to report that "80% of studies will yield a 95% confidence interval no wider than a given value," by taking account of the sampling distribution of the standard deviations.

The sampling distribution of the sample variances follows a chi-squared distribution. For a given tolerance level, the Tolerance Factor in the equation for confidence intervals is modified to reflect the required tolerance level (say, 80% rather than 50%). Specifically, we compute FactorSD = Sqrt(CHI_{Req} / df). This value is given by the function DCHIIN for 1 – Tolerance level, and df.

T-Test for Means (Two-Sample) with Estimated Variance

The width of the confidence interval computed for a completed study depends on the confidence level, the standard deviation, and the sample size. The confidence interval for a planned study depends on these values and also on the sampling distribution of the standard deviation.

The pooled within-group variance SD_p is given by

SD_p = Sqrt((((N1 – 1) * SD1 * SD1 + (N2 – 1) * SD2 * SD2) / (N1 + N2 – 2)))

The standard error of the mean difference is given by

SE_{diff} = SD_p * Sqrt(1 / N1 + 1 / N2)

The degree of freedom is given by

df = N1 + N2 – 2

The t-value corresponding to the confidence level (t_{CI}) is given by the central t-distribution for (1 – CI_{Level}) / Tails and df. The value is computed using the DTIN algorithm.

The lower and upper limits of the confidence interval are given by

LowerLimit = Diff − t_{CI} * SE * 50% Tolerance Factor
UpperLimit = Diff + t_{CI} * SE * 50% Tolerance Factor

where

the 50% Tolerance Factor = Sqrt(CHI_{Req} / df) for df=N1+N2−2. This value is given by the function DCHIIN for 1−Tolerance level, and df. This takes account of the sampling distribution of the standard deviation.

The value computed in this way is the median value. Half of the studies will yield a confidence interval wider than the expected value, and roughly half will yield a narrower interval.

One-Tailed versus Two-Tailed Tests

Assuming the 95% confidence level, the t-value computed for a one-tailed test would be based on 1−0.95 (that is, 0.05). The t-value corresponding to a two-tailed test would be based on (1-0.95)/2 (that is, 0.025).

The one-tailed "interval" yields a lower (or upper) limit closer to the observed value than does the two-tailed interval, but only one of the limits is meaningful. The one-tailed "interval" extends from minus infinity to the upper limit reported, or from the lower limit reported to positive infinity.

Tolerance Intervals

The program is also able to compute the tolerance interval for a given confidence interval—that is, to report that 80% of studies will yield a 95% confidence interval no wider than a given value—by taking account of the sampling distribution of the standard deviations.

The sampling distribution of the sample variances follows a chi-squared distribution. For a given tolerance level, the Tolerance Factor in the equation for confidence intervals is modified to reflect the required tolerance level (say, 80% rather than 50%). Specifically, we compute FactorSD = Sqrt(CHI_{Req} / df). This value is given by the function DCHIIN for 1−Tolerance level, and df.

Z-Test for Means (One-Sample)

The width of the confidence interval computed for a completed study depends on the confidence level, the standard deviation, and the sample size. The confidence interval width for a planned study depends on these values as well. Since the (known) standard deviation will be used in computing confidence intervals, the standard deviation observed in the sample has no impact on the width of the interval.

The z-value corresponding to the confidence level (Z_{CI}) is given by the normal distribution for $(1 - CI_{Level})$ / Tails. The value is computed using the DNORIN algorithm.

The standard error of the mean is given by SE = SD1 / Sqrt(N).

The expected value of the lower and upper limits of the confidence interval are given by

LowerLimit = Mean1 – Z_{CI} * SE
UpperLimit = Mean1 + Z_{CI} * SE

Since the standard deviation is known, rather than estimated, the width of the confidence interval is completely determined by the sample size and will not vary from study to study.

One-Tailed versus Two-Tailed Tests

Assuming the 95% confidence level, the z-value computed for a one-tailed test would be based on 1–0.95 (that is, 0.05). The z-value corresponding to a two-tailed test would be based on (1–0.95)/2 (that is, 0.025).

The one-tailed "interval" yields a lower (or upper) limit closer to the observed value than does the two-tailed interval, but only one of the limits is meaningful. The one-tailed "interval" extends from minus infinity to the upper limit reported, or from the lower limit reported to positive infinity.

Tolerance intervals

When the standard deviation is known, the width of the confidence interval will not vary from study to study.

Z-Test for Means (Two-Sample)

The width of the confidence interval computed for a completed study depends on the confidence level, the standard deviation, and the sample size. The confidence interval width for a planned study depends on these values as well. Since the (known) standard deviation will be used in computing confidence intervals, the standard deviation observed in the sample has no impact on the width of the interval.

The z-value corresponding to the confidence level (Z_{CI}) is given by the normal distribution for $(1 - CI_{Level})$ / Tails. The value is computed using the DNORIN algorithm.

The pooled within-group variance SDp is given by

SDp = Sqrt((((N1 - 1) * SD1 * SD1 + (N2 - 1) * SD2 * SD2) / (N1 + N2 – 2)))

The standard error of the mean difference is given by

$SE_{Diff} = SDp * Sqrt(1 / N1 + 1 / N2)$

The degrees of freedom is given by

$df = N1 + N2 - 2$

The z-value corresponding to the confidence level (Z_{CI}) is given by the normal distribution for $(1 - CI_{Level})$ / Tails. The value is computed using the DNORIN algorithm.

The expected value of the lower and upper limits of the confidence interval are given by

$LowerLimit = Mean1 - Z_{CI} * SE_{Diff}$
$UpperLimit = Mean1 + Z_{CI} * SE_{Diff}$

Since the standard deviation is known, rather than estimated, the width of the confidence interval is completely determined by the sample size and will not vary from study to study.

One-Tailed versus Two-Tailed Tests

Assuming the 95% confidence level, the z-value computed for a one-tailed test would be based on $1 - 0.95$ (that is, 0.05). The z-value corresponding to a two-tailed test would be based on $(1 - 0.95)/2$ (that is, 0.025).

The one-tailed "interval" yields a lower (or upper) limit closer to the observed value than does the two-tailed interval, but only one of the limits is meaningful. The one-tailed "interval" extends from minus infinity to the upper limit reported, or from the lower limit reported to positive infinity.

Tolerance Intervals

When the standard deviation is known, the width of the confidence interval will not vary from study to study.

One Proportion—Normal Approximation

The lower limit is computed as follows

If P=0, the lower limit is 0
If P=1, the lower limit is $((1 - CI_{Level}) / Tails)^{(1/n)}$

Otherwise, the lower limit is

Lower = (P + (a / 2) - z * Sqrt(((P * Q) / n) + (a / (4 * n)))) / (1 + a)

where

$Q = 1 - P$
Z is the z-value corresponding to $(1 - CI_{Level})/Tails$
$a = z^2 / n$
N is N of cases

The upper limit is computed as follows

If P = 1, the upper limit is 1
If P = 0, the upper limit is $1 - (((1 - CI_{Level}) / Tails)^{(1/n)})$

Otherwise, the upper limit is

Upper = (P + (a / 2) + z * Sqrt(((P * Q) / n) + (a / (4 * n)))) / (1 + a)

where

$Q = 1 - P$
Z is the z-value corresponding to $(1 - CI_{Level})/Tails$
$a = z^2 / n$
N is the N of cases

The confidence interval width computed in this manner approximates the value expected over an infinite number of trials.

One Proportion—Exact (Binomial) Formula

The confidence interval for a single proportion depends on the confidence level and the sample size. It also depends on the proportion of successes observed in the study, since the standard error of a proportion varies as a function of the proportion.

To compute the expected width of the confidence interval, we need to take account of the likelihood of all possible outcomes, and the confidence interval associated with each of these.

For a study with N = 10, the possible outcomes are 0 successes, 1 success, ... 10 successes, and the likelihood of each is given by the binomial distribution. This value is given by the DBINPR function for N cases, k successes, and true proportion as specified by the alternate hypothesis.

For each possible outcome, the lower limit (and the upper limit) of the confidence interval that would be reported may be computed using the binomial distribution. These values are computed by the DBINES function for N cases, k successes, and the specified confidence level.

The program iterates through every possible outcome, multiplies the lower limit of the confidence interval by the likelihood that the given outcome will actually be observed in the study, and then sums these products over all possible outcomes to yield the expected lower limit. The process is repeated for the upper limit.

The confidence interval width computed in this manner gives the expected width over an infinite number of trials.

One-Tailed versus Two-Tailed Intervals

For a two-tailed confidence interval, the confidence level is used as given (for example, 95 for 95%). For a one-tailed confidence "interval," the value used for 95% is 100 − 2*(100-95). The one-tailed "interval" yields a lower (or upper) limit closer to the observed value than does the two-tailed interval, but only one of the limits is meaningful. The one-tailed "interval" extends from minus infinity to the upper limit reported, or from the lower limit reported to positive infinity.

Two Independent Proportions—Normal Approximations

Rate Difference

With Z = the z-value corresponding to $(1 - CI_{Level})/Tails$, the lower and upper limits of the rate difference are given by

Lower = Diff − Z * SE_{Diff}
Upper = Diff + Z * SE_{Diff}

where

SE_{Diff} = Sqrt(((RowPct(1, 1) * RowPct(1, 2)) / Cell(1, 3))
    + ((RowPct(2, 1) * RowPct(2, 2)) / Cell(2, 3)))

and RowPct(Row,Col) refers to the percent of the row.

Odds Ratio

The lower and upper limits of the log odds are computed as

Lower = (LOGODDS − z * SE$_{LOGODDS}$)
Upper = (LOGODDS + z * SE$_{LOGODDS}$)

where

SE$_{LOGODDS}$ = Sqrt(1 / (Cell(1, 1)) + 1 / (Cell(1, 2)) + 1 / (Cell(2, 1)) + 1 / (Cell(2, 2)))

The lower and upper limit of the odds ratio are then given by

Lower = Exp(Lower Limit Log Odds)
Upper = Exp(Upper Limit Log Odds)

Relative Risk

The lower and upper limits of the log of the relative risk are computed as

Lower= (LOGRR − z * SE$_{LOGRR}$)
Upper= (LOGRR + z * SE$_{LOGRR}$)

where

SE$_{LOGRR}$ = Sqrt((Cell(1, 2) / Cell(1, 1)) / (Cell(1, 1) + Cell(1, 2))
         + (Cell(2, 2) / Cell(2, 1)) / (Cell(2, 1) + Cell(2, 2)))

The lower and upper limit of the relative risk are then given by

Lower = Exp(Lower Limit Log RR)
Upper = Exp(Upper Limit Log RR)

Two Independent Proportions—Cornfield Method

The logic of the Cornfield method is as follows. The chi-square test is used to test the null hypothesis that the observed sample was drawn from a specific population. Typically, this is a population in which the proportion of positive cases is identical in the two rows (that is, the test of independence). However, this is not a requirement of the test.

The logic of the Cornfield method is to identify the first population that will yield a significant effect (with alpha = 1 − CI$_{Level}$). Since a 2 × 2 table has only one degree of freedom, all four cells are completely determined by any single cell. In this example, we use the upper left cell.

Beginning with the observed sample, we subtract one case from the upper left cell (and adjust the other three cells as required by the marginals) to yield a "population," and test the observed study against this population. If the p-value exceeds 0.05 (assuming a 95% confidence level), we subtract an additional case. The process is repeated until we find a population that yields a p-value of exactly 0.05 for the test (the number of cases in the upper left cell can be a fractional value, so that a p-value of exactly 0.05 can be obtained).

The population identified in this way defines the lower limit. The rate difference, the odds ratio, and the relative risk corresponding to this population are reported as the lower limit for the population. To find the upper limit for each statistic, the process is repeated.

The computational methods used for confidence intervals for a 2×2 test of proportions are typically used to analyze a completed study, where the sample values are known. Since the sample values (which affect the variance) will vary from one sample to the next and cannot be known when the study is being planned, the confidence interval width reported here is only an estimate of the width that will be obtained in any given study.

Correlations (One-Sample)

The expected confidence interval for a correlation (r) is based on the Fisher-Z transformation of r (denoted Z_r).

The width of the confidence interval computed for a completed study depends on the confidence level, the observed correlation, and the sample size. The expected value of the confidence interval for a planned study depends on these values and also on the sampling distribution of the correlation coefficient.

The z-value corresponding to the confidence level (Z_{CI}) is given by the normal distribution for $(1 - CI_{Level})$ / Tails. The value is computed using the DNORN algorithm.

The confidence interval for r is computed using the Fisher-Z transformation of r,

$Z_r = 0.5 * Log((1 + r) / (1 - r))$

The standard error of Zr is given by

$SE = Sqrt(1 / (N1 - 3))$

The expected value of the lower and upper limits of the confidence interval in Zr units are given by

$Z_{LOWER} = Z_r - Z_{CI} * SE$
$Z_{UPPER} = Z_r + Z_{CI} * SE$

The expected values in r units are then computed by transforming the z-values into r values by

$r_z = (c - 1) / (c + 1)$, where $c = \mathrm{Exp}(2 * Z)$

The computational methods used for confidence intervals for a single correlation is typically used to analyze a completed study, where the sample values are known. Since the sample values (which affect the variance) will vary from one sample to the next and cannot be known when the study is being planned, the confidence interval width reported here is only an estimate of the width that will be obtained in any given study.

One-Tailed versus Two-Tailed Tests

Assuming the 95% confidence level, the t-value computed for a one-tailed test would be based on 1 − 0.95 (that is, 0.05). The t-value corresponding to a two-tailed test would be based on (1 − 0.95)/2, (that is, 0.025).

The one-tailed interval yields a lower (or upper) limit closer to the observed value than does the two-tailed interval, but only one of the limits is meaningful. The one-tailed interval extends from minus infinity to the upper limit reported, or from the lower limit reported to positive infinity.

Glossary

alpha

Alpha is the criterion required to establish statistical significance. Assuming that the null hypothesis is true, alpha is also the proportion of studies expected to result in a type 1 error. See "tails."

beta

The proportion of studies that will yield a type 2 error. Equal to one minus power.

confidence interval

Interval that will include the population parameter in a known proportion of all possible studies.

confidence level

A level of certainty for the confidence interval; the proportion of studies in which the confidence interval is expected to include the population parameter.

effect size

The magnitude of the effect—for example, the standard difference in means (for a t-test). Power analysis works with the effect size, which is independent of sample size. Significance tests combine the observed effect with the sample size to yield a combined test statistic.

paired proportions

A paired analysis is used when cases in one group are somehow linked with cases in the other group (for example, the researcher recruits pairs of siblings and assigns one to either treatment, or patients are matched for severity of disease). When the outcome is a dichotomous classification (proportion), the study is described as a paired proportion (also known as McNemar's test).

paired t-test

A paired analysis is used when cases in one group are somehow linked with cases in the other group (for example, the researcher recruits pairs of siblings and assigns one to either treatment, or patients are matched for severity of disease). When the study employs two groups and the outcome is continuous, the appropriate analysis is a paired (dependent) t-test.

power

Power is the proportion of studies that will yield a statistically significant effect (assuming the effect size, sample size, and criterion alpha specified in the study design).

precision

Used in this program to refer to the width of the confidence interval.

p-value

Computed for a significance test, the p-value gives the proportion of cases (for a population in which the effect is null) that will yield a sample in which the effect is as large (or larger) than the observed effect.

tails

A one-tailed test is used when an effect in one direction would be meaningful, but an effect in the opposite direction would have the same practical impact as no effect. Only an effect in the expected direction is interpreted. A two-tailed test is used when an effect in either direction would be meaningful (even if the researcher expects the effect to fall in a specified direction).

tolerance interval

Proportion of studies in which the width of the confidence interval is no greater than a given value.

type 1 error

The error committed when the true effect is null but the study yields a significant p-value and leads the researcher (in error) to reject the null hypothesis. With alpha set at 0.05, a type 1 error would be expected in 5% of trials in which the null hypothesis is true. By definition, a type 1 error will occur in 0% of trials in which the null hypothesis is false.

type 2 error

The error committed when the true effect is not null but the study fails to yield a significant p-value and the researcher (in error) fails to reject the null hypothesis. If power is 80%, a type 2 error would be expected in 20% of trials (assuming that the null hypothesis is false, which is an assumption of the power analysis). The type 2 error rate is also known as beta.

Bibliography

Altman, D. G. (1980). Statistics and ethics in medical research: How large a sample? *British Medical Journal* 281: 1336–1338.

Altman, D. G., Gore, S. M., Gardner, M. J., and Pocock, S. J. (1983). Statistical guidelines for contributors to medical journals. *British Medical Journal* 286: 1489–1493.

Bailar, J. C., and Mosteller, F. (1988). Guidelines for statistical reporting in articles for medical journals: Amplifications and explanations. *Annals of Internal Medicine* 108: 266–273.

Bakan, D. (1966). The effect of significance in psychological research. *Psychological Bulletin* 66: 423–437.

Berry, G. (1986). Statistical significance and confidence intervals. *Medical Journal of Australia* 144: 618–619.

Birnbaum, A. (1961). Confidence curves: An omnibus technique for estimation and testing statistical hypotheses. *Journal of the American Statistical Association* 56: 246–249.

Blackwelder, W. C. (1982). "Proving the null hypothesis" in clinical trials. *Controlled Clinical Trials* 3: 345–353.

Borenstein, M., Cohen, J., Rothstein, H. R., Pollack, S., et al. (1990). Statistical power analysis for one-way analysis of variance: A computer program. *Behavior Research Methods, Instruments, and Computers.*

Borenstein, M., Cohen, J., Rothstein, H. R., Pollack, S., and Kane, J. M. (1992). A visual approach to statistical power analysis on the microcomputer. *Behavior Research Methods, Instruments and Computers* 24: 565–572.

Borenstein, M. (1994a). Planning for precision in survival studies. *Journal of Clinical Epidemiology.*

———. (1994b). The case for confidence intervals in controlled clinical trials. *Controlled Clinical Trials* 15: 411–428.

———. (1994c). A note on confidence intervals in medical research. *Psychopharmacology Bulletin.*

———. (1997). Hypothesis testing and effect size estimation in clinical trials. *Annals of Allergy, Asthma, and Immunology* 78: 5–16.

Braitman, L. E. (1988). Confidence intervals extract clinically useful information from data (editorial). *Annals of Internal Medicine* 108: 296–298.

Brewer, J. K. (1972). On the power of statistical tests in the American Educational Research Journal. *American Educational Research Journal* 9: 391–401.

Brewer, J. K., and Sindelar, P. T. (1988). Adequate sample size: A priori and post hoc considerations. *The Journal of Special Education* 21: 74–84.

Bristol, D. R. (1989). Sample sizes for constructing confidence intervals and testing hypotheses. *Statistics in Medicine* 8: 803–811.

Brown, J., and Hale, M. S. (1992). The power of statistical studies in consultation-liaison psychiatry. *Psychosomatics* 33: 437–443.

Bulpitt, C. J. (1987). Confidence intervals. *Lancet* 1: 494–497.
Chase, L. J., and Chase, R. B. (1976). A statistical power analysis of applied psychological research. *Journal of Applied Psychology* 61: 234–237.
Cobb, E. B. (1985). Planning research studies: An alternative to power analysis. *Nursing Research* 34: 386–388.
Cohen, J. (1962). The statistical power of abnormal-social psychological research: A review. *Journal of Abnormal and Social Psychology* 65(3): 145–153.
_____. (1965). Some statistical issues in psychological research. In *Handbook of Clinical Psychology*, ed. B. B. Wolman, 95–121. New York: McGraw-Hill.
_____. (1977). *Statistical power analysis for the behavioral sciences—Revised edition.* Hillsdale, N.J.: Lawrence Erlbaum Associates, Inc.
_____. (1988). *Statistical power analysis for the behavioral sciences.* 2nd ed. Hillsdale, N.J.: Lawrence Erlbaum Associates, Inc.
_____. (1990). Things I have learned (so far). *American Psychologist* 45: 1304–1312.
_____. (1992) A power primer. *Psychological Bulletin* 112: 155–159.
_____. (1994). The earth is round (p<.05). *American Psychologist* 49: 997–1003.
Cohen, J., and Cohen, P. (1983). *Applied multiple regression/correlation analysis for the behavioral sciences.* 2nd ed. Hillsdale, N.J.: Lawrence Erlbaum Associates, Inc.
Detsky, A. S., and Sackett, D. L. (1985). When was a negative clinical trial big enough? *Archives of Internal Medicine* 145: 709–712.
Dunlap, W. P. (1981). An interactive FORTRAN IV program for calculating power, sample size, or detectable differences in means. *Behavior Research Methods and Instrumentation* 13: 757–759.
Dupont, W. D., and Plummer, W. D. (1990). Power and sample size calculations: A review and computer program. *Controlled Clinical Trials* 11: 116–128.
Feinstein, A. R. (1975). The other side of "statistical significance": Alpha, beta, delta, and the calculation of sample size. *Clinical Pharmacology and Therapeutics* 18: 491–505.
_____. (1976). Clinical biostatistics XXXVII. Demeaned errors, confidence games, nonplused minuses, inefficient coefficients, and other statistical disruptions of scientific communication. *Clinical Pharmacology and Therapeutics* 20: 617–631.
Fisher, R. A. (1955). Statistical methods and scientific induction. *Journal of the Royal Statistical Society, Series B* 17: 69–78.
Fleiss, J. (1979). Confidence intervals for the odds ratio in case-control studies: The state of the art. *Journal of Chronic Diseases* 32: 69–77.
_____. (1981). *Statistical methods in rates and proportions.* 2nd ed. New York: John Wiley and Sons, Inc.
Fleiss, J. L. (1986a). Confidence intervals vs. significance tests: Quantitative interpretation (letter). *American Journal of Public Health* 76: 587–588.
_____. (1986b). Significance tests have a role in epidemiologic research: Reactions to A. M. Walker. *American Journal of Public Health* 76: 559–560.
Foster, D. A., and Sullivan, K. M. (1987). Computer program produces p-value graphics. *American Journal of Public Health* 77: 880–881.
Freiman, J. A., Chalmers, T. C., Smith, H. J., and Kuebler, R. R. (1978). The importance of beta, the type II error and sample size in the design and interpretation of the randomized control trial: Survey of 71 "negative" trials. *New England Journal of Medicine* 299: 690–694.
Gardner, M. J., and Altman, D. G. (1986). Confidence intervals rather than P values: Estimation rather than hypothesis testing. *British Medical Journal (Clinical Research Edition)* 292: 746–750.

Gardner, M. J., and Altman, D. G. (1989a). *Statistics with confidence—Confidence intervals and statistical guidelines.* London: BMJ.
———. (1989b). *Statistics with confidence.* Belfast: The Universities Press.
Gart, J. J. (1962). Approximate confidence intervals for the relative risk. *Journal of the Royal Statistical Society* 24: 454–463.
Gart, J. J., and Thomas, D. G. (1972). Numerical results on approximate confidence limits for the odds ratio. *Journal of the Royal Statistical Society* 34: 441–447.
Gordon, I. (1987). Sample size estimation in occupational mortality studies with use of confidence interval theory. *American Journal of Epidemiology* 125: 158–162.
Gore, S. M. (1981). Assessing methods—Confidence intervals. *British Medical Journal* 283: 660–662.
Greenland, S. (1984). A counterexample to the test-based principle of setting confidence limits. *American Journal of Epidemiology* 120: 4–7.
———. (1988). On sample-size and power calculations for studies using confidence intervals. *American Journal of Epidemiology* 128: 231–237.
Hahn, G. J., and Meeker, W. Q. (1991). *Statistical intervals.* New York: John Wiley and Sons, Inc.
Hanley, J., and Lippman-Hand, A. (1983). "If nothing goes wrong, is everything all right?": Interpreting zero numerators. *Journal of the American Medical Association* 249: 1743–1745.
Hartung, J., Cottrell, J. E., and Giffen, J. P. (1983). Absence of evidence is not evidence of absence. *Anesthesiology* 58: 298–300.
Ingelfinger, F. J. (1975). The confusions and consolations of uncertainty (letter). *New England Journal of Medicine* 292: 1402–1403.
International Committee of Medical Journal Editors. (1991). Uniform requirements for manuscripts submitted to biomedical journals. *New England Journal of Medicine* 324: 424–428.
Kahn, H. A., and Sempos, C. T. (1989). *Statistical methods in epidemiology.* New York: Oxford University Press.
Kleinbaum, D. G., Kupper, L. L., and Morgenstern, H. (1982). *Epidemiologic research.* Belmont, Calif.: Lifetime Learning Publications.
Kraemer, H. C., and Thiemann, S. (1987). *How many subjects? Statistical power analysis in research.* Newbury Park: Sage.
Lachin, J. M. (1981). Introduction to sample size determination and power analysis for clinical trials. *Controlled Clinical Trials* 2: 93–113.
Machin, D., and Gardner, M. J. (1988). Reply to confidence intervals (letter). *British Medical Journal* 296: 1372.
Mainland, D. (1982). Medical statistics—Thinking vs. arithmetic. *Journal of Chronic Diseases* 35: 413–417.
Makuch, R. W., and Johnson, M. F. (1986). Some issues in the design and interpretation of "negative" clinical studies. *Archives of Internal Medicine* 146: 986–989.
McHugh, R. B., and Le, C. T. (1984). Confidence estimation and the size of a clinical trial. *Controlled Clinical Trials* 5: 157–163.
Morgan, P. P. (1989). Confidence intervals: From statistical significance to clinical significance (editorial). *Canadian Medical Association Journal* 141: 881–883.
Murphy, K. R. (1990). If the null hypothesis is impossible, why test it? *American Psychologist* 45: 403–404.

Nelson, N., Rosenthal, R., and Rosnow, R. L. (1986). Interpretation of significance levels and effect sizes by psychological researchers. *American Psychologist* 1299–1301.

Phillips, W. C., Scott, J. A., and Blaszczynski, G. (1983). The significance of "no significance": What a negative statistical test really means. *American Journal of Radiology* 141: 203–206.

Poole, C. (1987a). Mr. Poole's response (letter). *American Journal of Public Health* 77: 880.

_____. (1987b). Beyond the confidence interval. *American Journal of Public Health* 77: 195–199.

_____. (1987c). Confidence intervals exclude nothing. *American Journal of Public Health* 77: 492–493.

Reed, J. F., and Slaichert, W. (1981). Statistical proof in inconclusive "negative" trials. *Archives of Internal Medicine* 141: 1307–1310.

Reynolds, T. B. (1980). Type II error in clinical trials (editor's reply to letter). *Gastroenterology* 79: 180

Rosenthal, R., and Rubin, D. (1985). Statistical analysis: Summarizing evidence versus establishing facts.

Rosner, B. (1990). *Fundamentals of biostatistics.* Boston: PWS-Kent Publishing Co.

Rosnow, R. L., and Rosenthal, R. (1988). Focused tests of significance and effect size estimation in counseling psychology. *Journal of Counseling Psychology* 35: 203–208.

Rothman, K. J. (1978). A show of confidence (letter). *New England Journal of Medicine* 299: 1362–1363.

_____. (1986a). Significance questing (editorial). *Annals of Internal Medicine* 105: 445–447.

_____. (1986b). *Modern epidemiology.* Boston: Little, Brown and Company.

Rothman, K. J., and Yankauer, A. (1986). Editors' note. *American Journal of Public Health* 76: 587–588.

Rothstein, H. R., Borenstein, M., Cohen, J., and Pollack, S. (1990). Statistical power analysis for multiple regression/correlation: A computer program. *Educational and Psychological Measurement.*

Sedlmeyer, P., and Gigerenzer, G. (1989). Do studies of statistical power have an effect on the power of studies? *Psychological Bulletin* 105: 309–316.

Sheehe, P. R. (1993). A variation on a confidence interval theme (letter). *Epidemiology*, 4: 185–187.

Simon, R. (1986). Confidence intervals for reporting results of clinical trials. *Annals of Internal Medicine* 105: 429–435.

Smith, A. H., and Bates, M. N. (1992). Confidence limit analyses should replace power calculations in the interpretation of epidemiologic studies. *Epidemiology* 3: 449–452.

_____. (1993). A variation on a confidence interval theme (reply to letter). *Epidemiology* 4: 186–187.

Sullivan, K. M., and Foster, D. A. (1990). Use of the confidence interval function. *Epidemiology* 1: 39–42.

Thompson, W. D. (1987a). Exclusion and uncertainty (letter). *American Journal of Public Health* 77: 879–880.

_____. (1987b). Statistical criteria in the interpretation of epidemiologic data [published erratum appears in *American Journal of Public Health*, April 1987, 77(4): 515]. *American Journal of Public Health* 77: 191–194.

_____. (1993). Policy and p-values (letter) [published erratum appears in *American Journal of Public Health*, April 1987, 77(4): 515]. *American Journal of Public Health.*

Tiku, M. L. (1967). Tables of the power of the F-test. *Journal of the American Statistical Association* 62: 525–539.

Tversky, A., and Kahneman, D. (1971). Belief in the law of small numbers. *Psychological Bulletin* 76: 105–110.

Walker, A. M. (1986a). Significance tests represent consensus and standard practice (letter). *American Journal of Public Health* 76: 1033–1034.

———. (1986b). Reporting the results of epidemiologic studies. *American Journal of Public Health* 76: 556–558.

Wonnacott, T. (1985). Statistically significant. *Canadian Medical Association Journal* 133: 843.

Zelen, M., and Severo, N. C. (1964). Probability functions. In *Handbook of Mathematical Functions*, ed. M. Abromowitz, I. Stegan, et al., National Bureau of Standards. Washington, D.C.: U. S. Government Printing Office.

Index

Add/remove sets
  for multiple regression, 150
Alpha, 26
  Defined, 205
  In tables, 36
  Link to confidence level, 26
  Modify, 26
  Role in power analysis, 7
  Valid range, 26
Analysis of variance/ covariance (factorial), 127
Analysis of variance/ covariance (oneway), 113
Ancova (factorial)
  Algorithms, 190, 192
Ancova (oneway)
  Algorithms, 190, 192
Anova (factorial)
  Algorithms, 190
  Entering effect size directly, 131
  Entering effect size using conventions, 131
  Entering effect size using means, 134
  Entering effect size using range, 133
  Entering effect size using standard deviations, 132
  Entering effect size using variance, 132
  Selecting factors for graph, 143
  Selecting factors for table, 142
Anova (oneway)
  Algorithms, 190
  Entering effect size, 115
  Entering effect size directly, 117
  Entering effect size using conventions, 117
  Entering effect size using means, 120
  Entering effect size using range, 119
  Entering effect size using SD, 118
  Entering effect size using variances, 118
  Various methods for effect size, 120, 134
Arcsin method
  Algorithms (one-sample), 182
  Algorithms (two sample), 183
  Computational option (one sample), 77
  Computational options (two sample), 84

Beta
  Defined, 205
Binomial formula
  Proportions (one sample), 78

Casagrande and Pike, 83
Chi-Square test
  Algorithms (two-sample), 185
Clipboard
  Graphs, 47
  Main screen, 30
  Reports, 52
  Sensitivity analysis, 25
  Tables, 39
Computational algorithms for power, 177
Computational algorithms for precision, 195
Confidence interval
  Defined, 205
  Graphs, 45
Confidence level, 26
  Defined, 205
  In tables, 36
  Link to alpha, 26
  Modify, 26
  Role in precision, 12
  Valid range, 26
Conventions. See Effect size conventions
Correlation
  Pre/Post in paired t-test, 61
Correlation (one sample), 103
  Algorithms, 188
Correlation (two sample), 109
  Algorithms, 189
Covariates
  Anova (factorial), 138
  Anova (oneway), 123
  Multiple regression, 146
Crosstabulation. See KxC Crosstabulation

Index

Crosstabulation. See Paired proportions
Crosstabulation. See Proportions (one sample)
Crosstabulation. See Proportions (two sample)

d (effect size one-sample t-test), 54
Designated set
  in multiple regression, 147

Effect size
  ANCOVA (factorial), 138
  ANCOVA (oneway), 123
  ANOVA - Automatic update effect size, 121
  ANOVA (factorial)", 128
  ANOVA (oneway, 114
  ANOVA (Update automatically), 135
  Conventions, 22
  Correlation (one sample), 104
  Correlation (two sample), 110
  Crosstabulation (KxC), 98
  Defined, 205
  In tables, 35
  Multiple regression, 148
  Paired proportions, 88
  Proportions (one sample), 76
  Proportions (two sample), 82
  Role in power analysis, 5
  Role in precision, 14
  Sign test, 94
  t-test (one sample), 54
  t-test (paired), 61
  t-test (two sample), 68
Error model
  in multiple regression, 148
Ethical issues
  in power analysis, 10
Exact test (one sample proportions)
  Algorithms, 182

f (effect size for Anova -factorial), 128
f (effect size for Anova-oneway), 114
Factorial analysis of covariance, 138
Factorial Anova. See Anova (Factorial)
Files

Toolbar, 20
Find-N, 28
  Modify required power, 29
  N linked in two groups, 28
Fisher's exact test, 84
  Algorithms, 185

General case, 163
Graphs, 41
  Color, 47
  Confidence interval, 45
  Copy to clipboard, 47
  Customize, 45
  Default graph, 42
  Legends, 46
  Modify content, 45
  Modify format, 46
  Multiple regression sets, 151
  Power by alpha, 43
  Power by effect size, 43
  Precision, 45
  Print, 47
  Save, 48
  Selecting factors for Anova (factorial), 143
  Standard error, 44
  Titles, 46
Groups
  Link N in two groups, 28

Help
  Toolbar, 20

Kramer-Greenhouse, 84
KxC crosstabulation, 97
  Algorithms, 187

Large effect. See Effect size conventions
Link
  N in two groups, 28
  Standard deviations, 69

Main screen, 19
  Copy to clipboard, 30
  Effect size conventions, 22
  Sensitivity analysis, 23
  Summary Panel, 22
McNemar. See Paired proportions
Medium effect. See Effect size conventions
Model 1 error
  Algorithms, 193
  in multiple regression, 148
Model 2 error
  Algorithms, 193
  in multiple regression, 148
Modify
  Alpha, 26
  Confidence level, 26
  Tails, 26
Modify display
  Toolbar, 20
Multiple regression, 145
  Add/Remove sets, 150
  Algorithms, 192
  Designated set, 147
  Error model, 148
  Graphs, 151
  Polynomial regression, 157
  Tables, 151
  Templates, 150

Navigating
  Overview, 2
  Toolbar, 20
N-Cases, 28
  Link N in two groups, 28
  Set for tables, 38
  Spin control, 27
Nil hypothesis
  vs null hypothesis, 11
Non-central chi-square
  General case, 170
  Proportions (two sample), 84
Non-central F, 167
Non-central t, 164
Normal approximation
  Proportions (one sample), 78

Normal approximations (two sample)
  Algorithms, 184
Null hypothesis
  vs nil hypothesis, 11

Odds ratio, 84
One Tailed. See Tails
Oneway analysis of covariance. See Ancova (oneway)
Oneway analysis of variance. See Anova (oneway)
Oneway Anova. See Anova (oneway)
Overview of program, 1

Paired proportions
  Defined, 205
Paired proportions (McNemar), 87
  Algorithms, 186
Paired t-test. See t-test (paired)
Power
  Defined, 206
  Modify required power, 29
  Reset default for required power, 29
  Role in power analysis, 10
Precision
  Confidence level, 26
  Defined, 206
  Graphs, 45
  Planning for, 11
  Role of effect size, 14
  Role of sample size, 12
  Role of tails, 13
  Role of variance, 14
Printing
  Graphs, 47
  Reports, 51
  Sensitivity analysis, 25
  Tables, 39
Program overview, 1
Proportions (one sample), 75
  Algorithms, 182
  Binomial formula, 78
  Computational options, 78
  Normal approximation, 78
Proportions (two-sample)
  Computational options, 89

Proportions in two independent groups, 81
P-value
  Defined, 206

Regression. See Multiple Regression
Relative risk, 84
Reports, 49
  Copy to clipboard, 52
  Printing, 51
  Saving, 52
Retrieve data, 30

Sample size, 28
  Link N in two groups, 28
  Role in power analysis, 9
  Role in precision, 12
  Set for tables, 38
  Spin control, 27
Save
  Data, 30
  Graphs, 48
  Reports, 52
Saving
  Sensitivity analysis, 25
  Tables, 40
Selecting
  Statistical Procedures, 1
Sensitivity analysis
  Copy to clipboard, 25
  Printing, 25
  Saving, 25
  Sorting, 25
Sets (Add/Remove)
  for multiple regression, 150
Sign test, 93
  Algorithms, 187
  Computational options, 95
Small effect. See Effect size conventions
Sorting
  Sensitivity analysis, 25
Spin control
  N-cases, 27
Standard deviation
  Link in two groups, 69

t-test (one sample), 54
t-test (paired), 61
t-test (two sample), 68
Standard error
  Graphs, 44
Statistical procedures
  Ancova (factorial), 138
  Ancova (oneway), 123
  Anova/Ancova (factorial), 127
  Anova/Ancova (oneway), 113
  Correlation (one sample), 103
  Correlation (two sample), 109
  Examples, 1
  General case, 163
  KxC crosstabulation, 97
  Multiple regression, 145
  Paired proportions (McNemar), 87
  Proportions (one sample), 75
  Proportions in two independent groups, 81
  Selecting, 1
  Sign test, 93
  Summary, 1
  t-test (one sample), 53
  t-test (paired), 59
  t-test (two sample), 67
Summary Panel, 22

Tables, 31
  Copy to clipboard, 39
  Effect size, 35
  Formats, 32
  Graphing, 40
  Multiple regression sets, 151
  Power and precision, 32
  Power only, 32
  Printing, 39
  Saving, 40
  Selecting factors for Anova (factorial), 142
  Set sample size, 38
  Setting alpha, 36
  Setting confidence level, 36
  Setting tails, 36
  Style, 32
  Vary effect size and alpha, 37
Tails, 26
  Defined, 206
  In tables, 36

Modify, 26
Precision, 26
Role in power analysis, 9
Role in precision, 13
Valid range, 26
Templates
    for multiple regression, 150
Tolerance intervals, 16
    Defined, 206
    Role in precision, 16
    t-test (one sample), 56
Toolbar, 20
    Files, 20
    Help, 20
    Modify display, 20
    Navigating, 20
    Tools, 20
Tools
    Effect size conventions, 22
    Sensitivity analysis, 23
    Summary panel, 22
    Toolbar, 20
t-test
    Data entry options, 68
    vs z-test (one sample), 56
    vs z-test (paired), 64
    vs z-test (two sample), 71
t-test (one sample)
    Algorithms, 178
    Computational options (Power), 56
    Computational options (Precision), 56
    Variance known/estimated, 56
t-test (paired), 59
    Algorithms, 178
    Computational options (power), 64
    Computational options (precision, 64
    Defined, 206
    Variance known/estimated, 64
t-test (two sample), 67
    Algorithms, 179, 180, 181
    Computational options (power), 71
    Variance known/estimated, 71
Two tailed. See Tails
Type I error
    Defined, 207
Type II error
    Defined, 207

Variables (Add/Remove)
    for multiple regression, 150
Variance
    Known/estimated in t-test (one sample), 56
    Known/estimated in t-test (paired), 64
    Known/estimated in t-test (two sample), 71
    Role in precision, 14

Wizard. See Interactive wizard

Yates correction
    Proportions (two sample), 84

z-test vs t-test
    One sample, 56
    Paired, 64
    Two sample, 71